Selected Papers from the ISTEGIM'19

Selected Papers from the ISTEGIM'19—Thermal Effects in Gas Flow in Microscale

Editors

Lucien Baldas
Jürgen J. Brandner
Gian Luca Morini

MDPI • Basel • Beijing • Wuhan • Barcelona • Belgrade • Manchester • Tokyo • Cluj • Tianjin

Editors
Lucien Baldas
Université de Toulouse
France

Jürgen J. Brandner
Karlsruhe Institute of
Technology (KIT)
Germany

Gian Luca Morini
Alma Mater Studiorum
Università di Bologna
Italy

Editorial Office
MDPI
St. Alban-Anlage 66
4052 Basel, Switzerland

This is a reprint of articles from the Special Issue published online in the open access journal *Micromachines* (ISSN 2072-666X) (available at: https://www.mdpi.com/journal/micromachines/special_issues/ISTEGIM_19).

For citation purposes, cite each article independently as indicated on the article page online and as indicated below:

LastName, A.A.; LastName, B.B.; LastName, C.C. Article Title. *Journal Name* **Year**, *Volume Number*, Page Range.

ISBN 978-3-0365-0100-0 (Hbk)
ISBN 978-3-0365-0101-7 (PDF)

Cover image courtesy of Jürgen J. Brandner.

© 2021 by the authors. Articles in this book are Open Access and distributed under the Creative Commons Attribution (CC BY) license, which allows users to download, copy and build upon published articles, as long as the author and publisher are properly credited, which ensures maximum dissemination and a wider impact of our publications.
The book as a whole is distributed by MDPI under the terms and conditions of the Creative Commons license CC BY-NC-ND.

Contents

About the Editors vii

Lucien Baldas, Jürgen J. Brandner and Gian Luca Morini
Editorial for the Special Issue "Selected Papers from the ISTEGIM'19—Thermal Effects in Gas Flow in Microscale"
Reprinted from: *Micromachines* **2020**, *11*, 879, doi:10.3390/mi11090879 1

Shahin Mohammad Nejad, Silvia Nedea, Arjan Frijns and David Smeulders
The Influence of Gas–Wall and Gas–Gas Interactions on the Accommodation Coefficients for Rarefied Gases: A Molecular Dynamics Study
Reprinted from: *Micromachines* **2020**, *11*, 319, doi:10.3390/mi11030319 5

Zhenying Wang, Daniel Orejon, Khellil Sefiane and Yasuyuki Takata
Effect of Substrate Conductivity on the Transient Thermal Transport of Hygroscopic Droplets during Vapor Absorption
Reprinted from: *Micromachines* **2020**, *11*, 193, doi:10.3390/mi11020193 17

Marcel Nachtmann, Julian Deuerling and Matthias Rädle
Molecule Sensitive Optical Imaging and Monitoring Techniques—A Review of Applications in Micro-Process Engineering
Reprinted from: *Micromachines* **2020**, *11*, 353, doi:10.3390/mi11040353 27

Dominique Fratantonio, Marcos Rojas-Cárdenas, Christine Barrot, Lucien Baldas and Stéphane Colin
Velocity Measurements in Channel Gas Flows in the Slip Regime by means of Molecular Tagging Velocimetry
Reprinted from: *Micromachines* **2020**, *11*, 374, doi:10.3390/mi11040374 43

Hiroki Yamaguchi and Kenji Kito
Measurement of Heat Transfer from Anodic Oxide Film on Aluminum in High Knudsen Number Flows
Reprinted from: *Micromachines* **2020**, *11*, 234, doi:10.3390/mi11030234 75

Salvina Gagliano, Giovanna Stella, Maide Bucolo
Real-Time Detection of Slug Velocity in Microchannels
Reprinted from: *Micromachines* **2020**, *11*, 241, doi:10.3390/mi11030241 83

Danish Rehman, Jojomon Joseph, Gian Luca Morini, Michel Delanaye and Juergen Brandner
A Hybrid Numerical Methodology Based on CFD and Porous Medium for Thermal Performance Evaluation of Gas to Gas Micro Heat Exchanger
Reprinted from: *Micromachines* **2020**, *11*, 218, doi:10.3390/mi11020218 97

Jojomon Joseph, Danish Rehman, Michel Delanaye, Gian Luca Morini, Rabia Nacereddine, Jan G. Korvink and Juergen J. Brandner
Numerical and Experimental Study of Microchannel Performance on Flow Maldistribution
Reprinted from: *Micromachines* **2020**, *11*, 323, doi:10.3390/mi11030323 113

Natheer Almtireen, Jürgen J. Brandner and Jan G. Korvink
Numerical Thermal Analysis and 2-D CFD Evaluation Model for An Ideal Cryogenic Regenerator
Reprinted from: *Micromachines* **2020**, *11*, 361, doi:10.3390/mi11040361 131

Anna G. Ciriolo, Rebeca Martínez Vázquez, Alice Roversi, Aldo Frezzotti, Caterina Vozzi, Roberto Osellame and Salvatore Stagira
Femtosecond Laser-Micromachining of Glass Micro-Chip for High Order Harmonic Generation in Gases
Reprinted from: *Micromachines* **2020**, *11*, 165, doi:10.3390/mi11020165 . . . . **145**

Daniel Mariuta, Arumugam Govindaraji, Stéphane Colin, Christine Barrot, Stéphane Le Calvé, Jan G. Korvink, Lucien Baldas and Jürgen J. Brandner
Optofluidic Formaldehyde Sensing: Towards On-Chip Integration
Reprinted from: *Micromachines* **2020**, *11*, 673, doi:10.3390/mi11070673 . . . . **155**

About the Editors

Lucien Baldas (Prof. Dr.) is an expert in the numerical and experimental analysis of gas microflows, including heat transfer aspects. From 2008 to 2012, he was the assistant coordinator of the GASMEMS European Network (ITN 215504; www.gasmems.eu), which is aimed at structuring European research in the field of gas microflows, and led a work package in the MIGRATE European Project (ITN-ETN n° 643095), a research and training network on MIniaturized Gas flow for Applications with enhanced Thermal Effects. He has co-authored more than 100 papers in international journals and international conferences in the field of microfluidics.

Jürgen J. Brandner (Prof. Dr.) has about 25 years of expertise in microstructure devices for process engineering and miniaturized heat exchangers, materials and manufacturing for microscale systems, microsensors and sensor implementation, as well as measurement technologies for microscale devices, including in-situ and operando measurement with optical and electro-magnetic analysis systems such as high-speed videography or NMR spectroscopy. He was a member of the GASMEMS European Network (ITN 215504, www.gasmems.eu) and coordinated the MIGRATE European Project (ITN 643095, www.migrate2015.eu) from 2015 to 2019. He has authored or co-authored more than 350 papers in international journals and international conferences in the field of micro process engineering, microfluidics, micro manufacturing and heat transfer. He is a member of the Editorial Board of *Micromachines* (MDPI).

Gian Luca Morini (Prof. Dr.) is a Full Professor of Applied Thermal Engineering at the Department of Industrial Engineering (DIN) of the University of Bologna. He is the author of about 200 technical papers, which have appeared in the most important international journals and conferences related to microfluidics, applied thermal engineering, heat transfer and fluid flows. He is an Italian delegate of the EUROTHERM Committee, an Italian delegate of the Scientific Council of International Center of Heat and Mass Transfer (ICHMT), and a member of the Assembly of the World Conference (AWC) on Experimental Heat Transfer, Fluid Mechanics, and Thermodynamics. He is an Associate Editor of the *International Journal of Thermal Sciences* (Elsevier) and *Micromachines* (MDPI).

Editorial

Editorial for the Special Issue "Selected Papers from the ISTEGIM'19—Thermal Effects in Gas Flow in Microscale"

Lucien Baldas [1],*, Jürgen J. Brandner [2],* and Gian Luca Morini [3],*

1. Institut Clément Ader (ICA), Université de Toulouse, CNRS-INSA-ISAE-Mines Albi-UPS, 31400 Toulouse, France
2. Institute of Microstructure Technology, Campus Nord, Hermann-von-Helmholtz-Platz 1, 76344 Eggenstein-Leopoldshafen, Germany
3. Microfluidics Laboratory, Department of Industrial Engineering (DIN), University of Bologna, Via del Lazzaretto 15/5, 40131 Bologna, Italy
* Correspondence: lucien.baldas@insa-toulouse.fr (L.B.); juergen.brandner@kit.edu (J.J.B.); gianluca.morini3@unibo.it (G.L.M.)

Received: 10 September 2020; Accepted: 18 September 2020; Published: 21 September 2020

MIGRATE (www.migrate2015.eu) was an H2020 Marie Skłodowska-Curie European Training Network running from November 2015 to October 2019, intended to address some of the current challenges to innovation that face European industry with regard to heat and mass transfer in gas-based micro-scale processes. This project had 10 participants and 6 associated partners coming from all over the European community. It covered different aspects of enhanced heat transfer and thermal effects in gases: from modeling of heat transfer processes and devices, development and characterization of sensors and measurement systems for heat transfer in gas flows, and thermally-driven micro gas separators, to micro-scale devices for enhanced and efficient heat recovery in environmental applications, transport, telecommunications and energy generation. In the framework of this MIGRATE project, a 2-day symposium, ISTEGIM'19–Thermal Effects in Gas flow in Microscale, was held on 24–25 October 2019, in Ettlingen, Germany, with the aim to showcase the main achievements of the project and to exchange information on the covered topics with the wider scientific community. This Special Issue collects 11 papers (10 research articles and one review) presented during the symposium by members of the MIGRATE Network, by invited speakers and by contributing authors.

Two of these papers deal with fundamental issues encountered in microfluidic systems involving gases. Understanding gas–solid surface interactions under rarefied conditions is of primary importance for efficient modeling of gas flows and heat transfer in microsystems. Mohammad Nejad et al. [1] used a molecular dynamics approach to simulate the behavior of two monoatomic gases, argon and helium, confined between two gold surfaces, kept at different temperatures in order to determine the corresponding energy and momentum accommodation coefficients. They showed that the pressure level has a minor influence on the accommodation coefficients value. On the contrary, by comparing the simulation results with empirical and numerical values in the literature, the gas–solid interatomic potential based on ab-initio calculations was shown to be more reliable for computing accommodation coefficients than the Lorentz–Berthelot and Fender–Halsey mixing rules. In liquid desiccant dehumidification systems, the thermal effect induced by interfacial phase change affects the spatio-temporal evolution of the mass flux at the liquid–air interface. Wang et al. [2] investigated experimentally, using optical imaging and infrared (IR) thermography, the thermal effects during vapor absorption into hygroscopic liquid desiccant droplets. They showed that the substrate conductivity strongly affects the transient heat transfer process. In addition, the temperature change during vapor absorption was shown to be much lower than that induced by the concentration change, which permits the possible application of a simplified isothermal model to capture the main mechanisms during vapor absorption into hygroscopic droplets.

Local experimental data are of crucial importance for the characterization of liquid and gaseous flows in microchannels. As highlighted in the review by Nachtmann et al. [3], molecule-sensitive, spatially-resolved technologies (e.g., Raman spectroscopy or fluorescence techniques) which combine spatial and temporal resolution with molecular selectivity and no disturbance of fluidics, can be efficiently applied for monitoring and measuring in microchannels. Advantages and disadvantages of the current state-of-the-art approaches are analyzed in that review and guidelines are proposed for the selection of a suitable measuring system according the application. In Fratantonio et al. [4] molecular tagging velocimetry (MTV) by direct phosphorescence, another non-invasive optical measurement technique, is applied to argon and helium rarefied flows in a millimetric channel, providing the first flow visualizations ever reported of a confined gas flow in the slip regime (for Knudsen numbers down to 0.014). Despite the new challenges encountered in carrying out local velocity measurements in gas flows characterized by high molecular diffusion, this work demonstrated that MTV is currently the most promising technique for providing direct measurements of the slip velocity at the wall in rarefied gas flows. Gas–surface interaction plays an important role in the heat transfer from a hot to a cold surface in rarefied conditions. Yamaguchi and Kito [5] measured, in the free-molecular to near free-molecular flow regimes, heat fluxes from anodic oxide films on aluminum with different anodizing times through a gas confined between two surfaces with different temperatures, to extract the energy accommodation coefficient (EAC). They showed that the EAC increased with the surface roughness, which depends on the duration of the anodization process and which was observed by a scanning electron microscope. Velocity measurements of two-phase flows in microchannels, which play an important role in the design of portable devices for biological and chemical samples analysis are also an open issue due to the lack of low-cost detection systems. An approach based on a low-cost optical signal monitoring setup was used by Gagliano et al. [6] to investigate the slug flow generated by the interaction of two immiscible fluids (air and water) in two serpentine microchannels with diameters of 320 and 640 µm. The optical signal obtained from a photo-diode set-up was analyzed in real-time to obtain water and air slug velocity and frequency. The successful implementation of this real-time detection platform in LabVIEW was demonstrated, paving the way to the possible development of non-invasive, low-cost portable systems for micro-flow analysis which could also be suitable for an easy on-chip integration.

Heat transfer management is one of the major applications of fluidic microsystems. The numerical design of micro heat exchangers is a process consuming large amounts of computational resources due to the complexity of the device geometry (the inlet and outlet manifolds and the large number of parallel microchannels). Rehman et al. [7] proposed a novel hybrid numerical methodology in which the microchannels are modeled as a porous medium where a compressible gas is used as the working fluid. This reduced order model, based on a modified formulation of the Darcy–Forchheimer law to consider the compressible nature of gas flow in microchannels, was used to develop a complete model of a double layer micro heat exchanger with collecting and distributing manifolds and was validated by comparison with experimental data. The adopted hybrid methodology resulted in a savings of at least 20 million computational nodes compared to the meshing strategy adopted in a classic conjugate heat transfer numerical analysis. The high performances of micro heat exchangers in terms of heat transfer rates are due to the high surface to volume ratios linked to the small channel dimensions, but can also be improved by the generation of local turbulences in the microchannels thanks to perturbators. Joseph et al. [8] used a reduced order numerical model based on a porous medium approximation for the microchannels, to analyze flow maldistribution and pressure losses for two different perturbators: wire-net and S-shape perturbators. The wire-net microchannels showed a better performance in terms of thermal efficiency, while the S-shape fins provided better performance in terms of pressure loss. Experimental tests were performed on a single microchannel with S-shape perturbators and a micro heat exchanger with triangular collectors for the validation of the numerical simulations; the results showed that the perturbators provoke an earlier laminar to turbulent flow transition. In the last few decades, miniaturized regenerative cryocoolers found applications in various domains such as IR imaging systems, aerospace applications and cryosurgery, to cite a few. In these closed-cycle cryocoolers, the regenerator is a key component for the overall performance of the system. Almtireen et al. [9] developed a one-dimensional numerical model, based on the discretization of the ideal regenerator thermal equations, to study the thermal interaction between the working gas and the metallic porous medium for various geometric and operating conditions. A 2D axisymmetric Computational Fluid Dynamics (CFD) model was also built on ANSYS Fluent to evaluate the validity of the ideal regenerator model. Even if the 1D ideal model was capable of providing rather good predictions of the thermal interactions between the gas and the matrix material, it was shown that more accurate results provided by a 2D CFD model or a non-ideal 1D model are required for an efficient design or for the optimization of the regenerator element.

The last series of papers published in this Special Issue are devoted to more complex microsystems integrating multiple components on the same chip, for different types of applications. High-order harmonic generation (HHG) beams produced by ultrashort laser pulses focused in a gaseous medium are commonly exploited in the fields of extreme ultraviolet spectroscopy and attosecond science. However, due to their technological complexity, these HHG-based coherent light sources stayed confined, up to now, within a few advanced laboratories. In this context, Ciriolo et al. [10] presented the application of femtosecond laser micromachined complex glass microchips to the generation of high-order harmonics in gas. In their device, a microchannel filled with gas through a 3D network of gas distribution microchannels acts as a hollow waveguide for the driving laser. A considerable increase in harmonics generation efficiency was observed, compared with standard harmonic generation in gas jets; this opens new perspectives for the downscaling of an entire HHG-based beamline to a glass chip with numerous additional functionalities. There is currently a high demand for compact, portable, accurate and rapid gas detectors for indoor air monitoring. In this framework, Mariuta et al. [11] worked on the development of a miniaturized detector of formaldehyde (HCHO), which is one of the major indoor airborne pollutants. The proposed sensor is based on the Hantzsch reaction which occurs after HCHO is trapped in the liquid phase, coupled to the fluorescence optical sensing methodology. On-chip integration of the detection part of the sensor was presented in this paper. A fluorescence detector, based on an LED-induced fluorescence technique coupled to a CMOS image sensor as an ultra-low light detection system, was developed and proved to be capable of detecting 10 µg/L of formaldehyde derivatized into 3,5–diacetyl-1,4-dihydrolutidine (DDL) in a 3.5 µL interrogation volume.

A large part of the results reported in this Special Issue was obtained in the framework of the PhD theses of the students enrolled in the MIGRATE European Training Network. Other works presented in this issue also come from research groups in touch with the partners of the MIGRATE network. The brilliant results obtained by the MIGRATE early stage researchers during their work within a structured European training network demonstrate that, with efficient international cooperation among research centers and companies active in microfluidics, it becomes possible to enhance the research quality and accelerate the transition between fundamental research and applications. In addition, the network acts as catalyst of fruitful links with new partners on common topics. We think that the papers collected in this Special Issue demonstrate the strength of the Marie Skłodowska-Curie Actions promoted by the European Commission.

We wish to thank all authors who participated to the ISTEGIM'19 symposium and submitted their papers to this Special Issue. We would also like to acknowledge all the reviewers for dedicating their time to provide careful and timely reviews to ensure the quality of this Special Issue.

Funding: The ISTEGIM'19 symposium received funding from the European Community's Horizon 2020 Framework Programme under the Marie Sklodowska-Curie Grant Agreement No. 643095.

Conflicts of Interest: The authors declare no conflict of interest.

References

1. Mohammad Nejad, S.; Nedea, S.; Frijns, A.; Smeulders, D. The influence of gas–wall and gas–gas interactions on the accommodation coefficients for rarefied gases: A molecular dynamics study. *Micromachines* **2020**, *11*, 319. [CrossRef] [PubMed]
2. Wang, Z.; Orejon, D.; Sefiane, K.; Takata, Y. Effect of substrate conductivity on the transient thermal transport of hygroscopic droplets during vapor absorption. *Micromachines* **2020**, *11*, 193. [CrossRef] [PubMed]
3. Nachtmann, M.; Deuerling, J.; Rädle, M. Molecule sensitive optical imaging and monitoring techniques—A review of applications in micro-process engineering. *Micromachines* **2020**, *11*, 353. [CrossRef] [PubMed]
4. Fratantonio, D.; Rojas-Cárdenas, M.; Barrot, C.; Baldas, L.; Colin, S. Velocity measurements in channel gas flows in the slip regime by means of molecular tagging velocimetry. *Micromachines* **2020**, *11*, 374. [CrossRef] [PubMed]
5. Yamaguchi, H.; Kito, K. Measurement of heat transfer from anodic oxide film on aluminum in high knudsen number flows. *Micromachines* **2020**, *11*, 234. [CrossRef] [PubMed]
6. Gagliano, S.; Stella, G.; Bucolo, M. Real-time detection of slug velocity in microchannels. *Micromachines* **2020**, *11*, 241. [CrossRef] [PubMed]
7. Rehman, D.; Joseph, J.; Morini, G.L.; Delanaye, M.; Brandner, J. A hybrid numerical methodology based on CFD and porous medium for thermal performance evaluation of gas to gas micro heat exchanger. *Micromachines* **2020**, *11*, 218. [CrossRef] [PubMed]
8. Joseph, J.; Rehman, D.; Delanaye, M.; Morini, G.L.; Nacereddine, R.; Korvink, J.G.; Brandner, J.J. Numerical and experimental study of microchannel performance on flow maldistribution. *Micromachines* **2020**, *11*, 323. [CrossRef] [PubMed]
9. Almtireen, N.; Brandner, J.J.; Korvink, J.G. Numerical Thermal Analysis and 2-D CFD Evaluation model for an ideal cryogenic regenerator. *Micromachines* **2020**, *11*, 361. [CrossRef] [PubMed]
10. Ciriolo, A.G.; Vázquez, R.M.; Roversi, A.; Frezzotti, A.; Vozzi, C.; Osellame, R.; Stagira, S. Femtosecond laser-micromachining of glass micro-chip for high order harmonic generation in gases. *Micromachines* **2020**, *11*, 165. [CrossRef]
11. Mariuta, D.; Govindaraji, A.; Colin, S.; Barrot, C.; Le Calvé, S.; Korvink, J.G.; Baldas, L.; Brandner, J.J. Optofluidic formaldehyde sensing: Towards on-chip integration. *Micromachines* **2020**, *11*, 673. [CrossRef]

© 2020 by the authors. Licensee MDPI, Basel, Switzerland. This article is an open access article distributed under the terms and conditions of the Creative Commons Attribution (CC BY) license (http://creativecommons.org/licenses/by/4.0/).

Article

The Influence of Gas–Wall and Gas–Gas Interactions on the Accommodation Coefficients for Rarefied Gases: A Molecular Dynamics Study

Shahin Mohammad Nejad, Silvia Nedea, Arjan Frijns * and David Smeulders

Energy Technology, Department of Mechanical Engineering, Eindhoven University of Technology, 5600 MB Eindhoven, The Netherlands; s.mohammad.nejad@tue.nl (S.M.N.); s.v.nedea@tue.nl (S.N.); d.m.j.smeulders@tue.nl (D.S.)
* Correspondence: a.j.h.frijns@tue.nl

Received: 25 February 2020; Accepted: 17 March 2020; Published: 19 March 2020

Abstract: Molecular dynamics (MD) simulations are conducted to determine energy and momentum accommodation coefficients at the interface between rarefied gas and solid walls. The MD simulation setup consists of two parallel walls, and of inert gas confined between them. Different mixing rules, as well as existing ab-initio computations combined with interatomic Lennard-Jones potentials were employed in MD simulations to investigate the corresponding effects of gas-surface interaction strength on accommodation coefficients for Argon and Helium gases on a gold surface. Comparing the obtained MD results for accommodation coefficients with empirical and numerical values in the literature revealed that the interaction potential based on ab-initio calculations is the most reliable one for computing accommodation coefficients. Finally, it is shown that gas–gas interactions in the two parallel walls approach led to an enhancement in computed accommodation coefficients compared to the molecular beam approach. The values for the two parallel walls approach are also closer to the experimental values.

Keywords: rarefied gas; accommodation coefficient; molecular dynamics (MD) simulation; Ar–Au interaction; He–Au interaction; mixing rules; ab-initio potentials

1. Introduction

Rarefied gas condition is encountered in a broad range of modern engineering applications; for example, in low pressure devices such as semiconductor manufacturing and spacecraft flying at high altitudes, as well as small-scale structures such as microelectronic devices and micro/nanoelectromechanical systems (M/NEMS) [1,2]. In all these applications to achieve the effective thermal management, a fundamental understanding of gas–surface interactions (GSI) is of paramount importance. The degree of rarefaction of a gas is quantified by Knudsen number (Kn = $\frac{\lambda}{L}$), where λ is the mean free path of the gas molecule and L is the characteristic length scale. It is known that a gas is regarded as rarefied if Kn > 0.01 [1]. Over the years, a wide range of experimental [3–5] and numerical studies [6–8] have been carried out, investigating rarefied gas-solid surface interactions. Due to the complexities involved in the instrumentation, an empirical study of GSI is a very challenging and time-consuming task. Regarding computational techniques, due to the noncontinuum gas behavior adjacent to the solid surface, common continuum approaches (Navier–Stokes equations) are not applicable to describe energy and momentum exchange at GSI. Herein, particle-based simulations methods such as MD simulations [9] are considered a promising candidate to study GSI. MD simulations can provide an atomistic-level understanding of the scattering dynamics of the gas molecules interacting with solid surfaces. Energy and momentum accommodation coefficients (E/MACs), which are the most relevant parameters involved in GSI models, describe the degree at which a gas attains its thermal or kinematic equilibrium with a surface while interacting with it. MD simulation is a very promising tool to determine different accommodation coefficients. These coefficients can be fed into semi-empirical GSI models such as Maxwell's model [10] or Cercignani–Lampis–Lord (CLL) model [11] that can be employed as boundary conditions for higher-scale simulation techniques such as Direct Simulation Monte Carlo (DSMC) [12], Lattice Boltzmann method (LBM) [13], and method of moments (MoM) [14] to describe heat and mass flow at macroscopic level under rarefied condition. In literature there are various numerical studies in which MD simulations are employed to determine accommodation coefficients for different gas-solid surface combinations [15–22]. The general objective of all such investigations is to find the correlation between the energy and momentum accommodation coefficients and input parameters such as the gas temperature or purity, gas molecular weight (MW), surface condition (i.e., surface roughness, cleanness, temperature and chemistry), as well as the gas–surface interaction strength. Briefly summarized, results in the literature reveal that E/MACs decrease by increasing the temperature and implicitly the kinetic energy of the molecules. Moreover, increasing the surface roughness, gas molecular mass, and gas–solid interaction strength lead to an increase in E/MACs. Gas molecules approaching a surface are sometimes trapped by the potential well and stager on the surface for a while as they are physically adsorbed. The gas molecules may escape the potential well after some residence time through which they lose sufficient amount of their thermal or kinetic energy such that they accommodate to the surface temperature at a higher degree (i.e., resulting in a higher accommodation coefficient). This phenomenon called trapping-desorption is more likely to happen at lower temperature, higher surface roughness and, higher gas MW, as well as stronger gas-solid interaction which at the end causes higher E/MACs at aforementioned conditions.

Due to the superposition of many factors affecting the phenomenon of gas-surface interaction, sometimes in literature for the same pair of gas–solid surface, a considerable discrepancy in the values of accommodation coefficients, obtained by different MD simulation approaches is found. As an example, for the Platinum–Argon combination the obtained values for the tangential momentum accommodation coefficient by Yamamoto et al. [23] and Hyakutake et al. [24] were 0.19 and 0.89, respectively. Accurate values of accommodation coefficients are essential for the better assessment of the overall transport properties of rarefied gases. Therefore, we will compare most common MD approaches and study the effect on the resulting thermal and momentum accommodation coefficients, and compare the results with experimental values.

In most previous MD simulations [19,21,22] to compute E/MACs, the molecular beam approach was employed, in which the gas molecule adjacent to the surface is assumed to interact only with the wall atoms, and its initial velocity is sampled from an equilibrium distribution at certain temperature. Such assumptions are valid for a highly-rarefied gases (i.e., Kn > 10). However, the condition is very different in most M/NEMS applications, where less degree of rarefaction is encountered (Kn < 1) [1]. In such systems, in the case of a temperature difference between gas and surface, gas molecules will experience non-equilibrium processes. Furthermore, both of gas–wall and gas–gas interactions are equally important.

In the present work, two parallel plates MD approach is applied to calculate E/MACs of noble gases (Ar and He) interacting with Gold (Au) surface. Firstly, the dependence of E/MACs on the gas pressure in the system was investigated, which to our best knowledge has not been studied by MD simulations, previously. The impact of gas-solid interatomic potential on E/MACs was also characterized. To do so, a pairwise Lennard-Jones (LJ) potential was considered at the solid-gas interface. The LJ potential parameters were computed using different approximation methods such as Lorentz–Berthelot (LB) and Fender–Halsey (FH) mixing rules, as well as taking form existing ab-initio calculations. It has been observed that an interaction potential based on quantum calculations such as ab-initio computations is the most reliable one for computing different ACs. Such a behavior has been reported by Mane et al. [21] and Daun et al. [25], where using molecular beam approach, they studied the interaction between monoatomic gases with aluminum and iron surfaces, respectively. At the end, to unravel the importance of including gas–gas interactions on the MD obtained accommodations coefficients, we compare our two parallel plates results with those obtained by the molecular beam approach.

2. Materials and Methods

2.1. Molecular Dynamics (MD) Simulations

The MD simulation setup considered in this work is a three-dimensional system, in which a monatomic gas is confined between two parallel walls (see Figure 1).

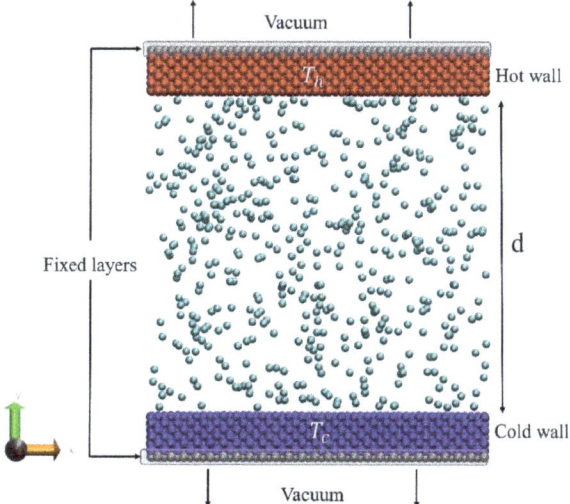

Figure 1. Schematic representation of the simulation setup; two walls kept at a distance d apart; they are thermostated at a low temperature T_c (bottom wall) and a high temperature T_h (top wall).

The cross-section area of the walls is 10 by 10 nm, and each of them consists of 8750 gold atoms arranged in a FCC structure. The walls are separated from each other in the y direction. The temperatures of the bottom and top walls are maintained at T_c = 300 K and T_h = 350 K, respectively, using Berendsen thermostats. In each wall, the outermost layer is fixed to prevent the wall from any translational or rotational motion. Periodic boundary conditions are applied in all three directions. Since the simulation box is also periodic in y direction, the walls are not placed directly at the periodic boundary. Vacuum has been considered between the walls at the periodic boundary in order to prevent direct contact and heat conduction between the hot and cold walls. At the beginning of the simulation, the temperature of the gas molecules is set at the 300 K, which corresponds to the root mean square velocity (V_{rms}) of 423 and 1367 m/s, for Ar and He gases, respectively. Herein, the gas molecules are not coupled to an external heat bath, and their temperature change is caused only via collisions with other particles (gas and solid particles). As it will be discussed in detail in the proceeding section, in order to obtain reliable results for E/MACs, a large number of collisions between gas molecules and the wall surface are required: 100,000 collisions have been considered here, as it has been recommended in previous studies [15,18] for a similar simulation setup. Increasing the number of collisions can be achieved by either extending the walls surface area or increasing the gas number density. Considering the computational time, the latter approach is more favorable. On the other hand, higher gas number density means higher pressure and the possibility to surpass the gas critical pressure, which is not desirable here, since the properties of a supercritical fluid differs in many aspects from a gas. In this work, in order to increase the gas number density, walls are positioned in the closest distance from each other in such a way that the pressure in the system does not exceed the critical pressure (P_{cr}) value in the system (For Ar: P_{cr} = 4.86 MPa, and for He: P_{cr} = 0.23 MPa [26]). This results in a separation distance of d = 11 nm and d = 102 nm for Ar and He gases, respectively. The Knudsen number in the case of the system with Ar (P_{Ar} = 2.75 MPa) and He (P_{He} = 0.21 MPa) gases are 0.23 and 0.56, respectively.

The interaction between Au atoms are described by the embedded atom model (EAM) potential [27], whereas the gas–gas and gas–solid interactions are modeled by the Lennard-Jones (LJ) potential. Herein, noble gas-Au pair potential coefficients have been calculated by commonly used (LB) mixing rule:

$$\sigma_{ij} = \frac{\sigma_{ii} + \sigma_{jj}}{2} , \; \varepsilon_{ij} = \sqrt{\varepsilon_{ii}\varepsilon_{jj}} \tag{1}$$

and (FH) mixing rule:

$$\sigma_{ij} = \frac{\sigma_{ii} + \sigma_{jj}}{2} , \varepsilon_{ij} = \frac{2\varepsilon_{ii}\varepsilon_{jj}}{\varepsilon_{ii} + \varepsilon_{jj}} \tag{2}$$

as well as existing ab-initio pair potentials [22,28], derived from computational quantum mechanics.

The detailed gas-gas and gases-Au interatomic potential parameters used for mixing rules are presented in Table 1. In addition, the pair potential coefficients based on existing ab-initio calculations are listed in Table 2. The LJ cut-off distance (r_c) for gas-gas interactions is set at 2.5 times the LJ length parameter (σ_{ii}). In addition, for both of Au-Ar and Au-He pairs we used r_c = 12 Å which is similar to the cut-off radii that were used in the ab-initio simulations [22].

Table 1. Gases-gold interaction potential parameters used by mixing rules and molecular weights.

Atom Type	ε_{ii} (meV)	σ_{ii} (Å)	MW (a.m.u)
Au [29]	229.4	2.63	196.96
Ar [30]	12.2	3.35	39.94
He [30]	0.94	2.64	4.00

Table 2. Gases-gold interaction potential parameters based on ab-initio computations [22].

Parameter	Value
ε_{Au-Ar}	11.36 (meV)
σ_{Au-Ar}	3.819 (Å)
ε_{Au-He}	0.787 (meV)
σ_{Au-He}	4.342 (Å)

For implementing MD simulations, LAMMPS molecular dynamics package was used [31]. All simulation setups were initially equilibrated for 1 ns (time step = 1 fs) for Ar and 3 ns for He. Afterwards, the production run was started and proceeded for the next 20 and 60 ns for Ar and He cases, respectively.

2.2. Computing Accommodation Coefficients

To compute E/MACs using MD simulations, the trajectory of each gas molecule in the simulation box is monitored during the specified simulation time. Herein, the collision is defined in such a way that if a gas molecule approaching the surface crosses a virtual border placed in certain distance from the surface, its velocity components are recorded. For the same particle after re-emitting from the surface, when it reaches back to the virtual border, its velocity components are recorded again. Both scattered and trapped particles that are desorbed from the wall are considered here. As it is depicted in Figure 2, in this study the virtual border is located at one gas-wall interaction cut off distance (r_c = 12 Å) away from the wall surface in order to guarantee that the gas molecule is not affected by the wall potential.

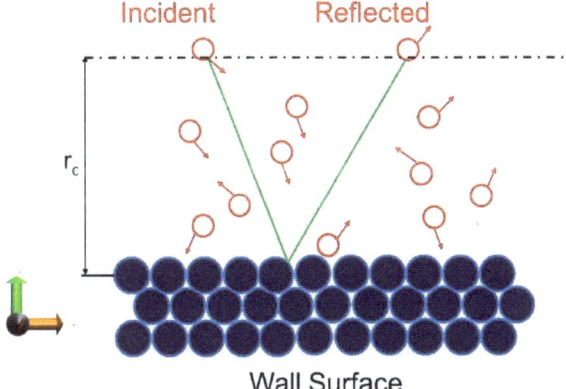

Figure 2. Schematic of gas-surface interaction used to compute accommodation coefficients in molecular dynamics (MD) simulations. Here the green line is a simplified example of gas molecule trajectory. The virtual border for the computation of the accommodation coefficients is placed at distance $r_c = 12$ Å from the solid surface.

The accommodation coefficients were calculated by the least-square method proposed by Spijker et al. [18]. In this approach, which is based on the correlation between input (Impinging) and output (Reflected) data obtained from MD simulations, the AC (α) can be computed as follows:

$$\alpha_q = 1 - \frac{\sum_i (Q_I^i - \langle Q_I \rangle)(Q_R^i - \langle Q_R \rangle)}{\sum_i (Q_I^i - \langle Q_I \rangle)^2}, \tag{3}$$

where subscript q can be any kinematic quantity, such as the gas molecule velocity in a certain direction (for momentum accommodation) or its total kinetic energy (for thermal energy). Q_I and Q_R referred to the considered quantity for the impinging and reflected particles, respectively. To be more specific, when a gas atom approaching to the surface ($v_y < 0$) passes through the virtual border (dotted line in Figure 2) it will classified as an impinging particle. From the other side, when a gas atom going outwards with respect to the surface ($v_y > 0$) passes through the virtual border it will be considered as reflected particle. The bracket notations denote that the average value for these quantities is calculated.

3. Results and Discussion

In order to compute E/MACs, after gathering collision data from MD simulation, the correlation between relevant impinging and outgoing velocities was studies. These correlations can be illustrated as a two-dimensional probability distribution profile, which for a particular impinging velocity gives the distribution of reflected velocities (see Figure 3). Herein, a line is fitted to the collision data based on least-square approximation (red line in Figure 3). Comparing the slope of this line, which is actually the fractional part of Equation (3), with dashed horizontal (fully diffusive) and diagonal (fully specular) lines gives us the accommodation coefficient.

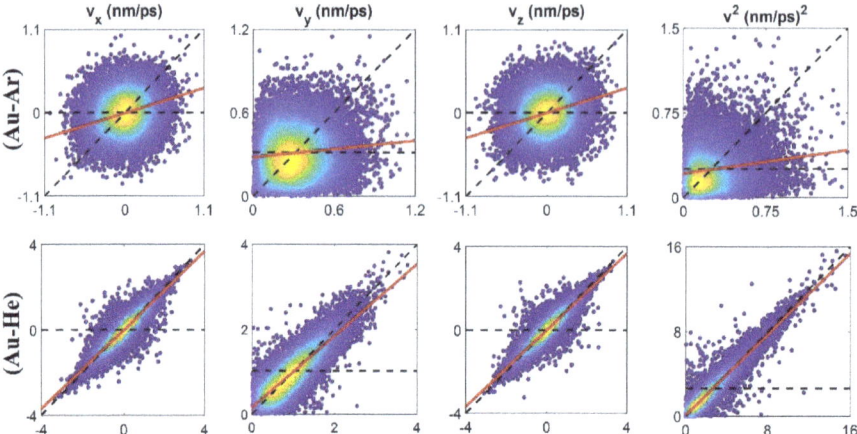

Figure 3. Velocity correlations of impinging (x-axis) and reflected (y-axis) velocity components of Ar and He on Au surface at 300 K using ab-initio potential. The dashes horizontal and diagonal lines indicate fully diffusive and specular conditions, respectively. The red line refers to the linear fit of the collision data obtained by MD simulations.

First of all, the dependency of computed E/MACs on gas pressure between two walls was investigated. Herein, ab-initio pair potential was employed to describe gas-wall interactions. By decreasing the number of gas atoms in the simulation box, the pressure was reduced from 2.75 MPa to 0.42 MPa, and from 0.21 to 0.04 MPa in the case of Ar and He, respectively. As it is depicted in Table 3, for both gases, decreasing pressure does not have a significant impact on obtained accommodation coefficients, but considerably increases the MD simulation time required to record the same number of collisions. Similar pressure dependency has been also reported in an experimental study by Thomas and Brown [32]. Therefore, in the remaining part of this paper we perform our simulations at the pressures of 2.75 and 0.21 MPa for Ar and He gases, respectively.

Table 3. Variation of energy and momentum accommodation coefficients and MD simulations running time with the pressure in the simulation box for Au–Ar and Au–He pairs.

System	Pressure (MPa)	Number Density (1/nm^3)	MFP (nm)	EAC	MAC	MD Simulations time (ns) *
Au–Ar	2.75	0.59	2.63	0.874	0.883	20
	1.27	0.27	5.71	0.832	0.846	50
	0.84	0.18	8.57	0.816	0.822	70
	0.42	0.09	17.14	0.783	0.791	100
Au–He	0.21	0.048	58.73	0.048	0.059	60
	0.13	0.029	97.89	0.046	0.057	90
	0.08	0.019	146.84	0.043	0.052	150
	0.04	0.009	293.70	0.042	0.049	250

* The time required to record 100,000 collisions

In the next step, the comparison between gas-wall interaction potentials obtained using different methods is shown in Figure 4. This figure shows that the FH mixing rule is relatively softer than the LB mixing, and that the ab-initio potential is softer than both mixing rules. Furthermore, it is depicted that the LB mixing rule highly overestimates the potential well depth. Such an overprediction in the case of a heavy gas like Ar causes that during MD run all Argon atoms are adsorbed on the solid surfaces (see Figure 5a), and that they do not leave the surface anymore. Therefore, for none of the impinging gas particles an outgoing velocity can be recorded, and E/MAC values are numerically unobtainable. In addition, the normalized number density distributions in the case of aforementioned system using varied interaction potentials are depicted in Figure 5b. Herein, initially it can be understood that the gas density adjacent to the wall surfaces is higher than the bulk density (n_0) in the central part of the system: stronger is the gas-wall interaction the higher is the density profile peak near to the wall. This is in agreement with the behavior reported in [33]. Furthermore, in the case of the LB potential, except in the vicinity of the walls, the gas density goes to zero.

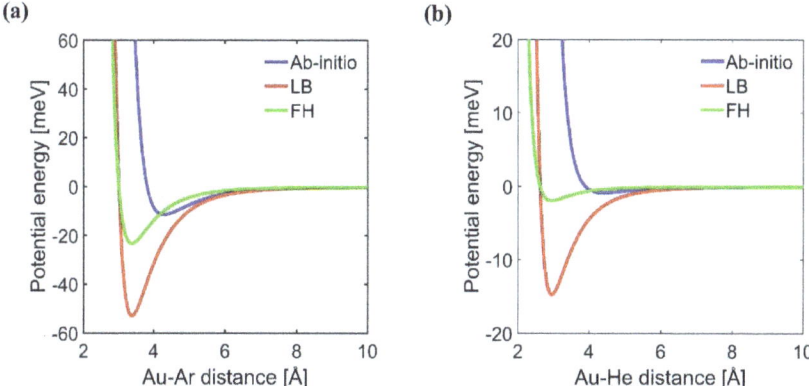

Figure 4. Pair potential energy plots of noble gases interaction with Au surface: (a) Au–Ar; (b) Au–He.

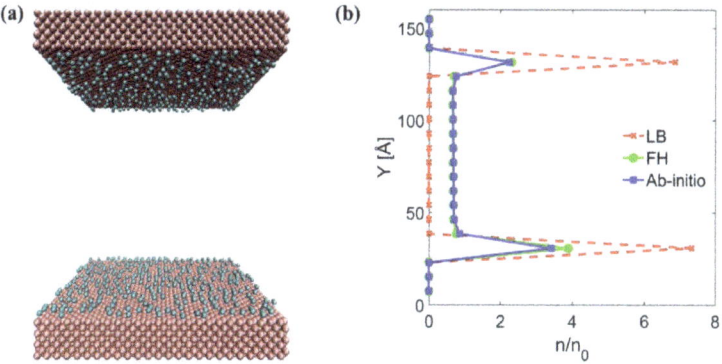

Figure 5. (a) Adsorption of Ar molecules on Au surface based on pair potential obtained from LB mixing rule; (b) normalized number density for different Au–Ar interaction potentials.

The obtained E/MACs for the aforementioned systems at the bottom wall ($T_w = 300$ K) are reported in Table 4. The reason why only E/MACs on bottom wall are reported here is for a further comparison with experimental data which are in the same temperature range. As it is shown in [15,18], the temperature gradient between two walls causes only a minor reduction in obtained values for E/MACs on the bottom wall. Therefore, this can be neglected. However, in order to resemble the experimental two parallel plates, in which the presence of a small temperature gradient ($T_h - T_c \ll T_c$) is a must [34], a temperature difference was imposed between the two plates in our MD simulations. For both noble gas cases, E/MACs increase with increasing the potential well depth. This is in agreement with the behavior as reported in other numerical studies [16,21]. The reason lies in the fact that higher gas-wall interaction strength increases the likelihood of trapping-desorption phenomenon, which at the end causes higher energy and momentum exchange between the solid surface and the neighboring gas.

In the case of Au–Ar, the reported empirical value for EAC (α_E) using two parallel plates approach at 296 K is 0.85 [34], which is in excellent agreement with obtained value for EAC using the potential based on ab-initio computations for our parallel walls assembly. Furthermore, Agrawal and Prabha [35] have reported that the tangential-MAC (TMAC) for Ar on commonly employed surface materials is 0.893, which this value also is consistent with obtained results for TMAC in different directions (α_x and α_z) in our case based on ab-initio pair potential. The EAC using FH mixing rule is 0.913, which is slightly higher than the value obtained by ab-initio potential.

Table 4. Momentum accommodation coefficient in three directions (α_x, α_y, α_z) and energy accommodation coefficients (α_E) results for Ar and He colliding with Au surface at T = 300 K using gas-wall interactions obtained different mixing rules, as well as existing ab-initio calculations.

System	Pair potential	α_x	α_y	α_z	α_E
Au–Ar	Ab-initio (Parallel walls)	0.824	0.913	0.832	0.874
	Ab-initio (Molecular beam) [22]	0.40	0.77	0.40	0.56
	Fender Halsey	0.915	0.934	0.913	0.913
	Experimental results: α_E = 0.85 [34]; TMAC(α_x, α_z) = 0.893 [35] *				
Au–He	Ab-initio (Parallel walls)	0.036	0.113	0.038	0.048
	Ab-initio (Molecular beam) [22]	0.013	0.046	0.014	0.017
	Fender Halsey	0.245	0.347	0.221	0.069
	Lorentz-Berthelot	0.642	0.748	0.653	0.187
	Experimental result: α_E = 0.31 [34]				

* No temperature is reported in reference [35].

For Au–He, Trott et al. [34] have measured EAC = 0.31 at 296 K, which is higher than all values reported in Table 4 for the same combination. To elucidate the possible reason behind the observed mismatch, it is noteworthy to mention that the surface roughness and the type of premeasurement treatment employed to clean the surface under investigation have significant impacts on obtained experimental results for E/MACs. For instance, for He–Pt pair at room temperature Mann [36] reported EAC = 0.03. For the same combination, in another experimental study [37] the measured EAC at 303 K was 0.238. The most important difference between the two aforementioned studies is that in the first study the metal surface is perfectly clean, but in the latter case the metal surface is somehow contaminated and it is also partially covered by testing gas. In addition, in another experiment [38] using Tungsten as the substrate, it was reported that EAC for He can vary from 0.017 (clean surface) to 0.23 (untreated surface). In Reference [34] it is also mentioned that recontamination of the surface is highly possible after the surface treatment technique that they used. Considering that in the performed MD simulation the surface is assumed to be atomistically smooth and clean, it can be inferred that MD results for E/MACs at the first place should be compared with experimental results on very clean and pure solid surface. Herein, assuming that clean Au surface has a similar behavior as clean Platinum and Tungsten surfaces, it is seen that the obtained results for EAC of Au–He using pair potential based on both FH mixing rule and ab-initio computations are similar to the values reported in [36] and [38] for He on clean Pt and Tungsten surface, respectively.

The presence of contamination on a real surface causes the accumulating of gas molecules on the surface which at the end leads to measuring higher E/MACs. To support this, the trajectories of Ar and He molecules interacting with Au surface using ab-initio computations pair potential during our MD simulation are depicted in Figure 6. In this figure the red crosses indicate the y-position of each gas molecules as function of simulation time. When an accumulation of red crosses is shown in certain area, it means that gas atom remains for a longer time in that area. It is shown that in the case of Ar molecule, which has higher molecular weight ($\frac{MW_{Ar}}{MW_{He}} \simeq 10$) and stronger interaction with Au surface, higher number of multiple collisions with Au surface has occurred (accumulation of red dots in vicinity of Au surface for Au–Ar pair). This means that in MD simulation of Au–Ar pair, there is higher chance that we see a layer of gas molecules adjacent to the solid surface. This layer of gas molecules can resemble the contamination in the case of experimental study, in which achieving a real clean surface is very challenging. Due to the presence of this layer on the surface in the case of Au–Ar combination, gas–gas interactions near the wall are more frequent. Since the gas–gas interaction strength is typically higher than gas-wall interaction strength (see Tables 1 and 2), it can be deduced that forming a gas layer at vicinity of surface leads to deriving higher values for E/MACs by MD simulations. Therefore, it is more likely that E/MACs obtained by MD simulations for a heavy argon gas match the experimental results.

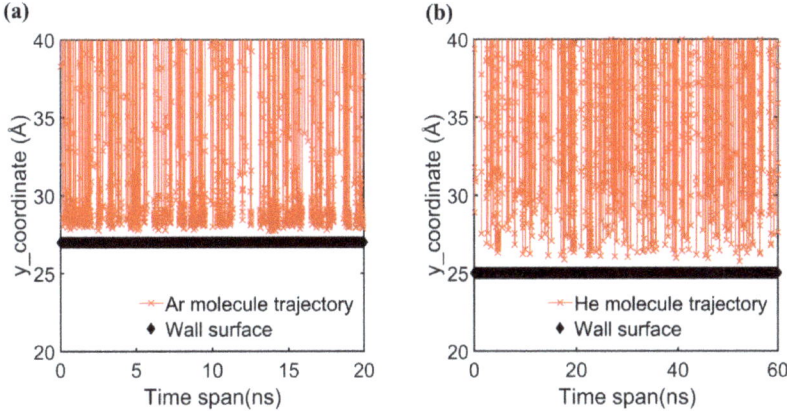

Figure 6. Trajectories of noble gases in vicinity of Au surface. (**a**) Ar–Au pair; (**b**) He–Au pair. Red crosses depict the y-coordinate of gas atoms as a function of simulation time.

Liao et al. [22] computed also E/MACs for Ar–Au and He–Au pairs using a molecular beam approach based on the same ab-initio potential that was employed here. As it is shown in Table 4, their results for E/MACs are lower than those obtained in this study. In the case of molecular beam approach gas molecules adjacent to the surface are considered to interact only with wall atoms, and gas-gas interactions are ignored. While in the case of the parallel plates approach due to presence of other gas molecules in the system, gas particles will be partly reflected back by other particles to the surface resulting in relatively more gas–wall interactions per molecule and therefore resulting in an increase in thermal and momentum accommodation coefficients.

4. Conclusions

The energy and momentum accommodation coefficients of monoatomic gases (Ar and He) on gold surface were determined through MD simulations. Initially, the impact of gas pressure on computed accommodation coefficients was investigated. Reducing gas pressure has a minor influence on accommodation coefficient values, whereas it can considerably increase the MD simulations time.

The effect of gas-wall interaction strength on accommodation coefficients was studied for different mixing rules, as well as for ab-initio pair potential. It was concluded that larger energy well depths in potential energy function leads to higher accommodation coefficients. For Ar–Au and He–Au gas-wall interactions, the energy well depth is overestimated for both Lorentz–Berthelot and for Fender–Halsey mixing rules resulting in an overestimation of the E/MACs. In the case of Ar–Au, the Lorentz–Berthelot mixing rule highly overestimates the potential well depth. This issue causes a fully saturated solid surface in which computing the accommodation coefficients from numerical point of view is not possible. It has been found out that the accommodation coefficients obtained by pair potential based on ab-initio computations are always in a reasonable agreement with experimental and numerical results in the relevant literature.

Comparing the obtained results for accommodation coefficients in this work with another study in which a molecular beam MD approach was used to compute accommodation coefficients reveals that gas–gas interaction is an important aspect that needs to be taken into account in the transient Knudsen regime since it leads to an enhancement in obtained accommodation coefficients.

Author Contributions: Conceptualization, S.M.N., S.N., and A.F.; methodology, S.M.N., S.N. and A.F.; validation, S.M.N.; investigation, S.M.N.; resources, A.F. and D.S.; writing—original draft preparation, S.M.N.; writing—review and editing, S.M.N., S.N. and A.F.; supervision, S.N., A.F. and D.S.; funding acquisition, A.F. All authors have read and agreed to the published version of the manuscript.

Funding: This work is part of the research program RareTrans with project number HTSM-15376, which is (partly) financed by the Netherlands Organization for Scientific Research (NWO).

Conflicts of Interest: The authors declare no conflict of interest.

References

1. Karniadakis, G.; Beskok, A.; Aluru, N. *Microflows and Nanoflows Fundamentals and Simulation*; Springer Science & Business Media: New York, NY, USA, 2006; Volume 29.
2. Rader, D.J.; Trott, W.M.; Torczynski, J.R.; Gallis, M.A.; Castañeda, J.N.; Grasser, T.W. *Microscale Rarefied Gas Dynamics and Surface Interactions for EUVL and MEMS Applications*; Department of Energy: Washington, DC, USA, 2004.
3. Saxena, S.C.; Joshi, R.K. *Thermal Accommodation and Adsorption Coefficients of Gases*; Hemisphere Publishing: New York, NY, USA, 1989.
4. Agrawal, A. A comprehensive review on gas flow in microchannels. *Int. J. Micro-Nano Scale Transp.* **2011**, *2*, 1–40. [CrossRef]
5. Cao, B.-Y.; Sun, J.; Chen, M.; Guo, Z.-Y. Molecular momentum transport at fluid-solid interfaces in MEMS/NEMS: A review. *Int. J. Mol. Sci.* **2009**, *10*, 4638–4706. [CrossRef] [PubMed]
6. Colin, S. Rarefaction and compressibility effects on steady and transient gas flows in microchannels. *Microfluid. Nanofluidics* **2005**, *1*, 268–279. [CrossRef]
7. Billing, G.D. The dynamics of molecule-surface interaction. *Comput. Phys. Rep.* **1990**, *12*, 383–450. [CrossRef]
8. Barker, J.A.; Auerbach, D.J. Gas-surface interactions and dynamics; Thermal energy atomic and molecular beam studies. *Surf. Sci. Rep.* **1984**, *4*, 1–99. [CrossRef]
9. Allen, M.P.; Tildesley, D.J. *Computer Simulation of Liquids*, 2nd ed.; Oxford University Press: Oxford, UK, 2017; Volume 53.
10. Maxwell, J.C., III. On stresses in rarefied gases arising from inequalities of temperature. *Proc. R. Soc. Lond.* **1878**, *27*, 304–308.
11. Lord, R.G. Some further extensions of the Cercignani-Lampis gas-surface interaction model. *Phys. Fluids* **1995**, *7*, 1159–1161. [CrossRef]
12. Bird, G.A. *Molecular Gas Dynamics and the Direct Simulation of Gas Flows*; Clarendon Press: Oxford, UK, 1994.
13. Zhang, J. Lattice Boltzmann method for microfluidics: Models and applications. *Microfluid. Nanofluidics* **2011**, *10*, 1–28. [CrossRef]
14. Grad, H. On the kinetic theory of rarefied gases. *Commun. Pure Appl. Math.* **1949**, *2*, 331–407. [CrossRef]
15. Prabha, S.K.; Sathian, S.P. Computational study of thermal dependence of accommodation coefficients in a nano-channel and the prediction of velocity profiles. *Comput. Fluids* **2012**, *68*, 47–53. [CrossRef]
16. Chirita, V.; Pailthorpe, B.A.; Collins, R.E. Non-equilibrium energy and momentum accommodation coefficients of Ar atoms scattered from Ni(001) in the thermal regime: A molecular dynamics study. *Nucl. Instrum. Methods Phys. Res. Sect. B Beam Interact. Mater. Atoms* **1997**, *129*, 465–473. [CrossRef]
17. Sun, J.; Li, Z.-X. Three-dimensional molecular dynamic study on accommodation coefficients in rough nanochannels. *Heat Transf. Eng.* **2011**, *32*, 658–666. [CrossRef]

18. Spijker, P.; Markvoort, A.J.; Nedea, S.V.; Hilbers, P.A.J. Computation of accommodation coefficients and the use of velocity correlation profiles in molecular dynamics simulations. *Phys. Rev. E* **2010**, *81*, 011203. [CrossRef] [PubMed]
19. Daun, K.J. Thermal accommodation coefficients between polyatomic gas molecules and soot in laser-induced incandescence experiments. *Int. J. Heat Mass Transf.* **2009**, *52*, 5081–5089. [CrossRef]
20. Reinhold, J.; Veltzke, T.; Wells, B.; Schneider, J.; Meierhofer, F.; Ciacchi, L.C.; Chaffee, A.; Thöming, J. Molecular dynamics simulations on scattering of single Ar, N_2, and CO_2 molecules on realistic surfaces. *Comput. Fluids* **2014**, *97*, 31–39. [CrossRef]
21. Mane, T.; Bhat, P.; Yang, V.; Sundaram, D.S. Energy accommodation under non-equilibrium conditions for aluminum-inert gas systems. *Surf. Sci.* **2018**, *677*, 135–148. [CrossRef]
22. Liao, M.; Grenier, R.; To, Q.-D.; de Lara-Castells, M.P.; Léonard, C. Helium and argon interactions with gold surfaces: Ab initio-assisted determination of the He–Au pairwise potential and its application to accommodation coefficient determination. *J. Phys. Chem. C* **2018**, *122*, 14606–14614. [CrossRef]
23. Yamamoto, K. Slightly rarefied gas flows over a smooth Pt surface. *AIP Conf. Proc.* **2001**, *585*, 339–346.
24. Hyakutake, T.; Yamamoto, K.; Takeuchi, H. Flow of gas mixtures through micro channel. *AIP Conf. Proc.* **2005**, *762*, 780–788.
25. Daun, K.J.; Sipkens, T.A.; Titantah, J.T.; Karttunen, M. Thermal accommodation coefficients for laser-induced incandescence sizing of metal nanoparticles in monatomic gases. *Appl. Phys. B* **2013**, *112*, 409–420. [CrossRef]
26. Cengel, Y.A.; Boles, M.A. *Thermodynamics: An Engineering Approach*, 8th ed.; McGraw-Hill Education: New York, NY, USA, 2015.
27. Sheng, H.W.; Kramer, M.J.; Cadien, A.; Fujita, T.; Chen, M.W. Highly optimized embedded-atom-method potentials for fourteen fcc metals. *Phys. Rev. B* **2011**, *83*, 134118. [CrossRef]
28. Grenier, R.; To, Q.-D.; de Lara-Castells, M.P.; Léonard, C. Argon interaction with gold surfaces: Ab initio-assisted determination of pair Ar–Au potentials for molecular dynamics simulations. *J. Phys. Chem. A* **2015**, *119*, 6897–6908. [CrossRef] [PubMed]
29. Heinz, H.; Vaia, R.A.; Farmer, B.L.; Naik, R.R. Accurate simulation of surfaces and interfaces of face-centered cubic metals using 12-6 and 9-6 lennard-jones potentials. *J. Phys. Chem. C* **2008**, *112*, 17281–17290. [CrossRef]
30. Schroeder, D.V. Interactive molecular dynamics. *Am. J. Phys.* **2015**, *83*, 210–218. [CrossRef]
31. Plimpton, S. Fast parallel algorithms for short-range molecular dynamics. *J. Comput. Phys.* **1995**, *117*, 1–19. [CrossRef]
32. Thomas, L.B.; Brown, R.E. The accommodation coefficients of gases on platinum as a function of pressure. *J. Chem. Phys.* **1950**, *18*, 1367–1372. [CrossRef]
33. Markvoort, A.J.; Hilbers, P.A.J.; Nedea, S.V. Molecular dynamics study of the influence of wall-gas interactions on heat flow in nanochannels. *Phys. Rev. E* **2005**, *71*, 066702. [CrossRef]
34. Trott, W.M.; Castaeda, J.N.; Torczynski, J.R.; Gallis, M.A.; Rader, D.J. An experimental assembly for precise measurement of thermal accommodation coefficients. *Rev. Sci. Instrum.* **2011**, *82*, 621. [CrossRef]
35. Agrawal, A.; Prabhu, S.V. Survey on measurement of tangential momentum accommodation coefficient. *J. Vac. Sci.* **2008**, *26*, 634–645. [CrossRef]
36. Mann, W. The exchange of energy between a platinum surface and gas molecules. *Proc. R. Soc. Lond. Ser. A Contain. Pap. A Math. Phys. Character* **1934**, *146*, 776–791.
37. Thomas, L.B.; Olmer, F. The accommodation coefficients of He, Ne, A, H2, D2, 02, C02, and Hg on platinum as a function of temperature. *J. Am. Chem. Soc.* **1943**, *65*, 1036–1043. [CrossRef]
38. Thomas, L.B.; Schofield, E.B. Thermal accommodation coefficient of helium on a bare tungsten surface. *J. Chem. Phys.* **1955**, *23*, 861–866. [CrossRef]

© 2020 by the authors. Licensee MDPI, Basel, Switzerland. This article is an open access article distributed under the terms and conditions of the Creative Commons Attribution (CC BY) license (http://creativecommons.org/licenses/by/4.0/).

Article

Effect of Substrate Conductivity on the Transient Thermal Transport of Hygroscopic Droplets during Vapor Absorption

Zhenying Wang [1,2,*], Daniel Orejon [1,3], Khellil Sefiane [1,3,4] and Yasuyuki Takata [1,2,*]

1. International Institute for Carbon-Neutral Energy Research (WPI-I2CNER), Kyushu University, 744 Motooka, Nishi-ku, Fukuoka 819-0395, Japan
2. Department of Mechanical Engineering, Thermofluid Physics Laboratory, Kyushu University, 744 Motooka, Nishi-ku, Fukuoka 819-0395, Japan
3. Institute for Multiscale Thermofluids, School of Engineering, The University of Edinburgh, Edinburgh EH9 3FD, Scotland, UK
4. Tianjin Key Lab of Refrigeration Technology, Tianjin University of Commerce, Tianjin 300134, China
* Correspondence: z.wang@heat.mech.kyushu-u.ac.jp (Z.W.); takata@mech.kyushu-u.ac.jp (Y.T.); Tel.: +81-092-802-3133 (Z.W.); +81-92-802-3100 (Y.T.)

Received: 25 January 2020; Accepted: 12 February 2020; Published: 13 February 2020

Abstract: In all kinds of liquid desiccant dehumidification systems, the temperature increase of the desiccant solution due to the effect of absorptive heating is one of the main reasons of performance deterioration. In this study, we look into the thermal effects during vapor absorption into single hygroscopic liquid desiccant droplets. Specifically, the effect of substrate conductivity on the transient heat and mass transfer process is analyzed in detail. The relative strength of the thermal effect and the solutal effect on the rate of vapor absorption is investigated and compared to the thermal effect by evaporative cooling taking place in pure water droplets. In the case of liquid desiccants, results indicate that the high thermal conductivity of copper substrates ensures more efficient heat removal, and the temperature at the droplet surface decreases more rapidly than that on Polytetrafluoroethylene (PTFE) substrates. As a result, the initial rate of vapor absorption on copper substrates slightly outweighs that on PTFE substrates. Further analysis by decomposing the vapor pressure difference indicates that the variation of vapor pressure caused by the temperature change during vapor absorption is much weaker than that induced by the concentration change. The conclusions demonstrate that a simplified isothermal model can be applied to capture the main mechanisms during vapor absorption into hygroscopic droplets even though it is evidenced to be unreliable for droplet evaporation.

Keywords: thermal effects; substrate conductivity; absorptive heating; evaporative cooling; vapor pressure difference

---

1. Introduction

Liquid desiccant is one type of aqueous salt solution characterized by its hygroscopic properties, and has been widely applied in various dehumidification and absorption systems [1,2]. Due to the existence of specific ions with strong adhesion to water molecules, the water vapor pressure at the droplet surface is reduced when compared to the partial vapor pressure of the surrounding air [3]. As a result, water vapor diffuses from the air side towards the liquid–air interface, and gets absorbed into the droplet [4]. Along with vapor–liquid phase change, the latent heat released will heat up the liquid solution, which is one of the main reasons of performance deterioration in all kinds of dehumidification devices [5].

Studies on the thermal effect taking place during the evaporation of sessile volatile droplets have been carried out extensively in the past years. The evaporation of volatile molecules cools down the droplet surface, and the effect of evaporative cooling is proved to strongly affect the evaporative mass flux. Typically, the thermal conductivity of the solid phase is several orders higher than that of the gas phase; therefore, heat conduction into the solid substrate plays an important role in the heat transfer process especially for droplets in still air with weak convection. Experiments carried out by Dunn et al. [6,7] confirm the strong effect of substrate conductivity on droplet evaporation, and an improved mathematical model is derived which relates the saturation vapor concentration at the droplet interface with the localized surface temperature. Sobac and Brutin [8] investigated the influence of substrate properties on the evaporation process in both hydrophilic and hydrophobic cases. Results highlight the need for more accurate models to account for the buoyant convection in vapor transport as well as the evaporative cooling and heat conduction between the droplet and the substrate [9]. Similar experiments were conducted by Talbot et al. [10] on picoliter droplets, which suggest that the thermal effects on the evaporation rate are much stronger for droplets on low-thermal-conductivity substrates than those on high-thermal-conductivity substrates. They also drew a similar conclusion that the evaporation time is underestimated by existing isothermal models.

To compensate the weakness of the isothermal models, Sefiane et al. [11] proposed a general expression for droplet evaporation which accounts for the thermal effect associated with evaporative cooling and includes the effects of both substrate and liquid properties. Similar theoretical trials were also conducted subsequently by Xu and Ma [12]. Zhang et al. [13] established a mathematical model to account for the thermal effect in an evaporating pure liquid droplet. The results show the nonmonotonic distribution of interfacial temperature, which is further explained combining the effect of evaporative cooling and heat of conduction through the liquid and the substrate. By solving a similar model using a finite element method, Wang et al. [14] characterized the combined effects of the underlying substrate and evaporative cooling. Results show that the influence of substrate properties on the evaporation process also depends on the strength of evaporative cooling. Other experimental and numerical investigations on the thermal effect also include the influence of substrate heating [15], thermal Marangoni [16], heat flux distribution [17], etc.

The existing studies indicate that the thermal effect induced by interfacial phase change affects the spatiotemporal evolution of mass flux at the liquid–air interface. Opposite to droplet evaporation, the vapor absorption into hygroscopic solution droplets will induce a strong effect of absorptive heating. The thermal effect along with heat conduction governs the temperature distribution within both the liquid droplet and the solid substrate, which in turn affects the rate of vapor absorption.

In our previous research, we investigated the mechanisms of droplet growth and spreading [4], as well as the effects of ambient temperature, humidity, and surface wettability on the vapor absorption process [18]. In this study, we investigate the thermal effects and demonstrate its relation with the substrate properties during vapor absorption into hygroscopic liquid desiccant droplets. Experiments are carried out for four representative environmental conditions, where the evolution of droplet profile and the temperature distribution at the droplet surface are extracted using optical imaging and infrared (IR) thermography. Results on substrates with different thermal conductivity and controlled wettability indicate the strong effect of substrate properties on the spatial-temporal evolution of interfacial temperature and mass flux. The relative strength of thermal effect on the transient heat transfer and on the air-side vapor diffusion during evaporation and vapor absorption are compared and summarized.

2. Materials and Methods

Experiments are conducted within an environmental chamber with accurately controlled conditions (800 L, −20–100 °C, 20–98% RH, PR-3KT from ESPEC Corp., Osaka, Japan). The accuracy of the temperature control is reported to be ± 0.5 °C, while the accuracy of the humidity control is ± 5% RH. Shown in Figure 1, during experiments, the evolution of droplet profile is recorded with a high-definition charge-coupled device (CCD) camera (Sentech STC-MC152USB with a RICOH lens and 25-mm spacing ring from OMRON SENTECH Corp., Kanagawa, Japan) at 4.8 fps, while a LED backlight is applied to enhance the image contrast. An IR camera, FLIR SC-4000 (Wilsonville, OR, USA), with a spectral range between 3.0 and 5.0 µm and a resolution of 18 mK, is set up vertically looking at the substrate and the deposited droplet from the top. The temperature evolution at the droplet liquid–gas interface is then recorded at 2 fps. Videos are subsequently processed with external software and self-developed programs, such as ImageJ® and Matlab®.

Polytetrafluoroethylene (PTFE) and copper, two types of commonly used packing materials in dehumidification systems, are applied as the testing substrates [1]. The dimensions of both copper and PTFE substrates are 20 mm × 20 mm × 10 mm, where the height is 10 mm. To rule out the influence of surface wettability on the droplet behavior, a uniform fluorinated ethylene propylene (FEP) coating layer is deposited onto both substrates following the same self-assembled monolayer (SAM) procedure. Since the thickness of the FEP coating is the same for both substrates and can be considered negligible when compared to the bulk material (thickness of the SAM is in the order of nanometers while the thickness of the studied samples is 10 mm), the effect of thermal conductivity of PTFE and copper can be investigated for droplets with similar contact angles of 106 ± 3°. 54 wt. % LiBr-H_2O solution from Sigma-Aldrich is used as the testing fluid for vapor absorption experiments, and the droplet volume is controlled as 2.5 ± 0.3 µL. Contrast experiments of droplet evaporation are conducted using distilled water (Sigma-Aldrich). Other detailed properties of the testing fluids and substrates are listed in Tables 1 and 2.

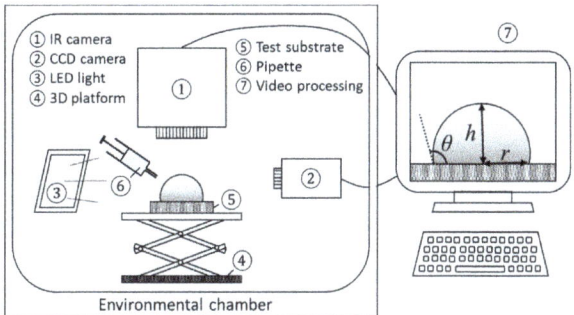

Figure 1. Overview of the experimental setup, including environmental chamber, charge-coupled device (CCD) camera, IR camera, back light, stainless steel vertical platform, droplet dosing system, and data acquisition system with ImageJ® and Matlab®.

Table 1. Properties of 54 wt. % LiBr solution and distilled water as: specific heat capacity c_p (kJ/kg/K); density ρ (kg/m³); liquid-gas surface tension γ_{lg} (mN/m); viscosity v (mPa·s); thermal conductivity k (W/m/K); and saturation temperature T_{sat} (°C). Properties shown were obtained at 20 °C and at 1 atm.

Liquid Type	c_p (kJ/kg/K)	ρ (kg/m³)	γ_{lg} (mN/m)	v (mPa·s)	k (W/m/K)	T_{sat} (°C)
54 wt.% LiBr solution	1.98	1600	91.54	4.751	0.4286	141
Distilled water	4.18	998	72.75	1.005	0.5984	100

Table 2. Properties of Polytetrafluoroethylene (PTFE) and copper substrates as density ρ (kg/m³); specific heat capacity c_p (kJ/kg/K); thermal conductivity k (W/m/K); thermal diffusivity α (m²/s), $\alpha = k/\rho c_p$; and thickness δ (mm) at 20 °C and 1 atm. We state here that thermal diffusivities α in this table are the right values compared to the values earlier reported [4,18].

Material	ρ (kg/m³)	c_p (kJ/kg/K)	k (W/m/K)	α (mm²/s)	δ (mm)
PTFE	2200	1.05	0.25	0.108	10.0
Copper	8960	0.39	397	114	10.0

Before experiments, the substrate samples are cleaned using an ultrasonic bath and deionized water, and are further dried with filtered compressed air to remove any possible remaining dusts or contaminants. After that, the testing fluid and substrate are placed inside the environmental chamber for more than 30 min until thermal-equilibrium state is attained. Then, a droplet with a controlled volume is deposited gently onto the substrate and then real-time recording of both CCD camera and IR camera is triggered. To ensure the reliability of the experimental results, each experiment is repeated 5 times. We note here that, since the characteristic length of the droplet lies below the capillary length ($\lambda = \sqrt{\gamma_{lg}/\rho g}$, ca. 2.7 mm for water and ca. 2.42 mm for 54 wt.% LiBr-H_2O solution at 20 °C), we assume the droplet shape as a spherical cap and derive the droplet volume and other parameters accordingly.

3. Results and Discussion

Figure 2 indicates the representative varying curves of contact angle θ and contact radius r of LiBr-H_2O droplets for 25 °C—60% RH and 45 °C—90% RH conditions, as well as the contrast pure water droplets for 25 °C—60% RH conditions on copper and PTFE substrates.

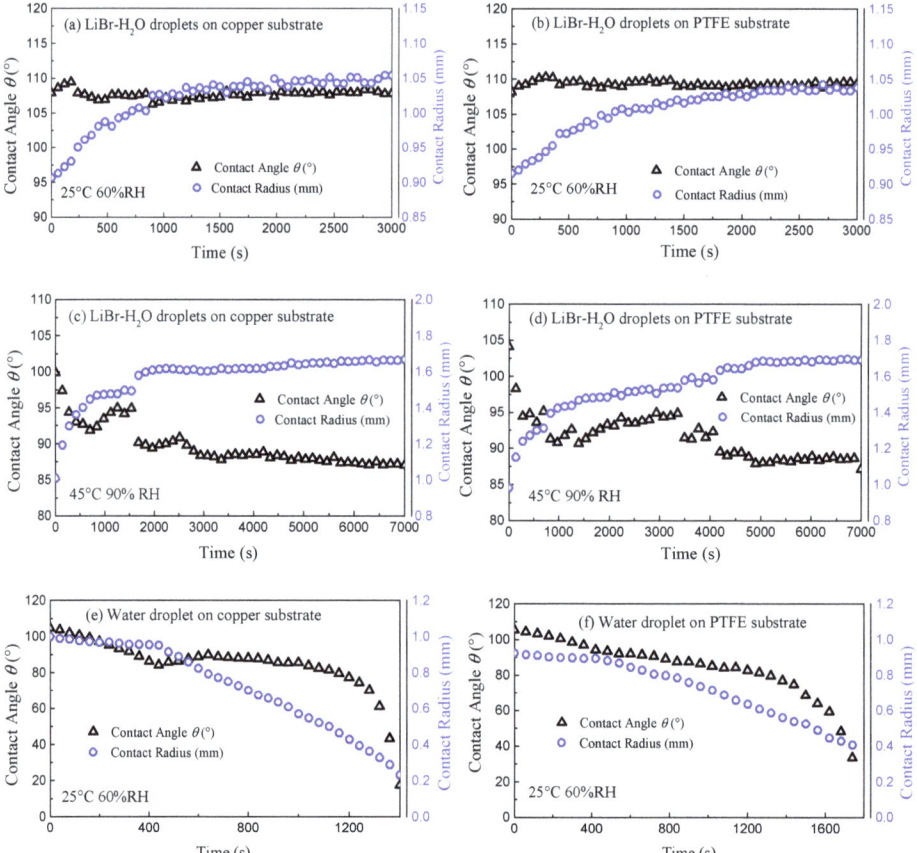

Figure 2. Evolution of contact angle θ (black triangles points) and contact radius r (blue circles points) of LiBr-H_2O (**a**–**d**) and pure water droplets (**e**,**f**) on copper (**a**,**c**,**e**) and PTFE (**b**,**d**,**f**) substrates coated with FEP for 25 °C—60% RH (**a**,**b**,**e**,**f**) and 45 °C—90% RH (**c**,**d**) conditions.

Due to the existence of uniform FEP coating, the evolution of contact angle and contact radius follows similar qualitative trend regardless of the substrate conductivity. For LiBr-H$_2$O droplets, at 25 °C and 60% RH, the contact angle increases slightly at first, then decreases until reaching fully equilibrium state, and stays constant during the remaining of the vapor absorption experiment. The initial increase of contact angle is due to fast vapor absorption, in which case the advancing contact line alone cannot keep up with the rapid volume expansion. As vapor absorption goes on, the solute concentration decreases due to water uptake, which causes the decrease in the liquid–gas surface tension and the rate of vapor absorption. As a joint result of decreasing surface tension and absorption rate, the contact angle decreases in the later stage. At the same time, the contact radius increases continuously following a saturation trend until the droplet reaches equilibrium with the ambient as presented in Figure 2a,b. In the case of 45 °C and 90% RH, upon deposition, the droplet contact angle is lower than that for 25 °C as a consequence of the lower surface tension of the liquid desiccant at higher temperature. Then, the contact angle decreases quickly at the initial moment. Along with vapor absorption, the contact angle fluctuates with the advancing stick-slip behaviors of the droplet (Figure 2c,d) [18] until both radius and contact angle show a plateau at which equilibrium is reached with the ambient.

For pure water droplets, the evolution of contact angle and contact radius follows the typical trend of an evaporating volatile droplet on a hydrophobic substrate [19,20]. At the initial period, the contact line remains pinned with decreasing contact angle, i.e., constant contact radius (CCR) mode. Then, the contact line starts to recede and the contact angle keeps constant, i.e., constant contact angle (CCA) mode. At the final stage, the small droplet diminishes with the contact angle and contact radius decreasing simultaneously as in the mixed mode.

Figure 3 shows the variation of normalized volume V/V_0 of LiBr-H$_2$O droplets and pure water droplets on copper and PTFE substrates taking 25 °C and 60% RH conditions as a representative example. At this ambient condition, the rate of vapor absorption is moderate while the vapor absorption phenomenon is apparent and the equilibrium state is easily reached. Due to vapor absorption, the volume of LiBr-H$_2$O droplets increases rapidly at first and then slows down as the droplet gets saturated with water. Comparing the increasing trend of the two curves, it can be seen that the rate of vapor absorption of droplets on copper substrates slightly outweighs that of droplets on PTFE substrates at the initial period (0–600s), while at the final equilibrium stage, the droplet volume attained in the two cases is the same. Differently, for pure water droplets, the evaporation rate is found to be greatly affected by the substrate conductivity. As shown in Figure 3b, water droplets on copper substrates exhibit an apparently higher evaporation rate and shorter lifetime than those on PTFE substrates, which corresponds with the results in previous studies [10,11].

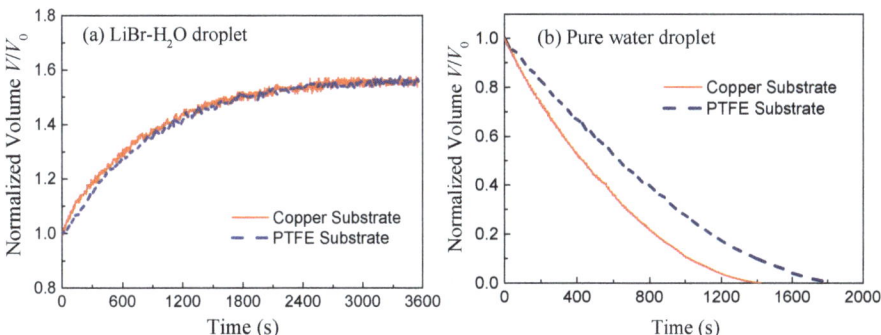

Figure 3. Evolution of normalized volume for (**a**) LiBr-H$_2$O droplets and (**b**) pure water droplets on copper (solid red lines) and PTFE (dashed blue lines) substrates coated with FEP along with time for 25 °C and 60% RH conditions.

Based on mass conservation, the solute concentration inside the LiBr-H$_2$O droplets (Figure 4a) can be calculated as established in the work of Wang et al. [4]. In addition, the vapor pressure difference between the ambient and the droplet surface (Fig. 4(b)) can be evaluated according to the fitting correlations derived by Patek et al. [21] and further implemented by Wang et al. for the absorption of LiBr-H$_2$O liquid desiccant droplets [4]:

$$\Delta P = P_{vapor,ambient} - P_{vapor,surface}, \quad P_{vapor,surface} = P_{sat}(\Theta), \qquad (1)$$

where $P_{vapor,ambient}$ is the partial vapor pressure of ambient air, $P_{vapor,surface}$ is the vapor pressure at the liquid–air interface, and P_{sat} is the saturation vapor pressure of pure water at shifted temperature, Θ. Θ is function of the mole fraction, x_{mole}, and temperature, T, of the LiBr-H$_2$O solution, and can be calculated as Equation (2):

$$\Theta = T - \sum_{i=1}^{8} a_i (x_{mole})^{m_i} |0.4 - x_{mole}|^{n_i} \left(\frac{T}{T_c}\right)^{t_i}, \qquad (2)$$

where T_c is the critical temperature of pure water, 647.096 K, x_{mole} is the mole fraction, $a = \{-2.41303 \times 10^2, 1.91750 \times 10^7, -1.75521 \times 10^8, 3.25432 \times 10^7, 3.92571 \times 10^2, -2.12626 \times 10^3, 1.85127 \times 10^8, 1.91216 \times 10^3\}$, $m = \{3, 4, 4, 8, 1, 1, 4, 6\}$, $n = \{0, 5, 6, 3, 0, 2, 6, 0\}$, and $t = \{0, 0, 0, 0, 1, 1, 1, 1\}$.

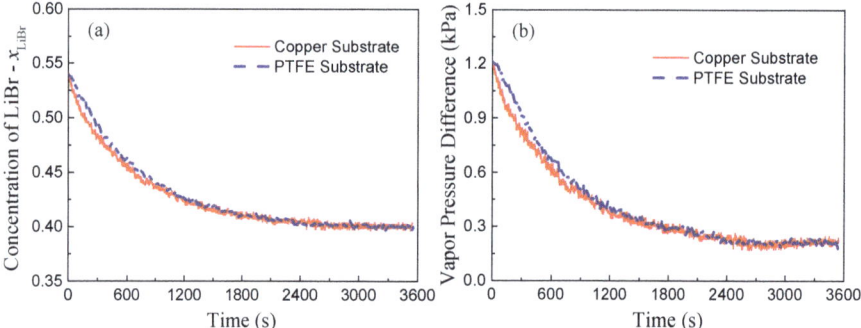

Figure 4. Evolution of (**a**) solute concentration, x_{LiBr}, and (**b**) vapor pressure difference between surrounding air and liquid-air interface of LiBr-H$_2$O droplets on copper and PTFE substrates coated with FEP for 25 °C and 60% RH conditions.

Calculation results indicate that the vapor pressure difference between the ambient and the droplet surface decreases along with time (Figure 4b), and since the driving force for vapor diffusion decreases, the rate of vapor absorption decreases accordingly. Different from other parameters, the substrate conductivity mainly affects the transient heat and mass transfer process. Therefore, despite the slight difference in the rate of vapor absorption during the initial non-equilibrium period, droplets on copper and PTFE substrates will finally reach the same state, i.e., temperature and solute concentration, with the same final volume for a given specific environmental condition; reaching thermal, chemical, and thermodynamic balance with the environment.

To provide more fundamentals and insights on the transient heat transfer during vapor absorption, the interfacial temperature of droplets on copper and PTFE substrates is investigated by IR thermography. Figure 5 shows the experimental results at 45 °C and 90% RH as the rate of vapor absorption is apparently high allowing for easier comparison. The spatial temperature distribution across the LiBr-H$_2$O droplet is overall homogenous throughout the vapor absorption process, indicating a negligible thermal Marangoni effect at the surface. The droplet surface experiences the highest temperature at the initial moment due to fast vapor absorption which starts right after the droplet is generated from the needle and contacts the humid air. After being deposited on the substrate, the absorbed heat is dissipated across the liquid droplet and into the solid substrate, and the interfacial temperature decreases along with time.

For droplets on the copper substrate (Figure 5a), it takes about 22 s for the interfacial temperature to decrease from the initial *ca.* 60 °C to *ca.* 52 °C, while on PTFE substrates (Figure 5b) it takes much longer, about 200 s from the initial *ca.* 59 °C to *ca.* 52 °C. The heat of conduction within the droplet and across the substrate can be evaluated by the characteristic time, expressed as, $\tau^* = \rho c_p h^2 / k$, where h is the characteristic length, i.e., the height of the droplet or the thickness of the substrate. By making use of the characteristic time τ^*, the time scale for heat conduction within the droplet is ~10 s, ~1 s for heat conduction across the copper substrate, and 10^2–10^3 s for heat conduction across the PTFE substrate. Results on the characteristic time for the heat of conduction agree with the experimental observations, and verify the dominating influence of substrate conductivity in the transient heat transfer and evolution of interfacial temperature of the droplet during vapor absorption.

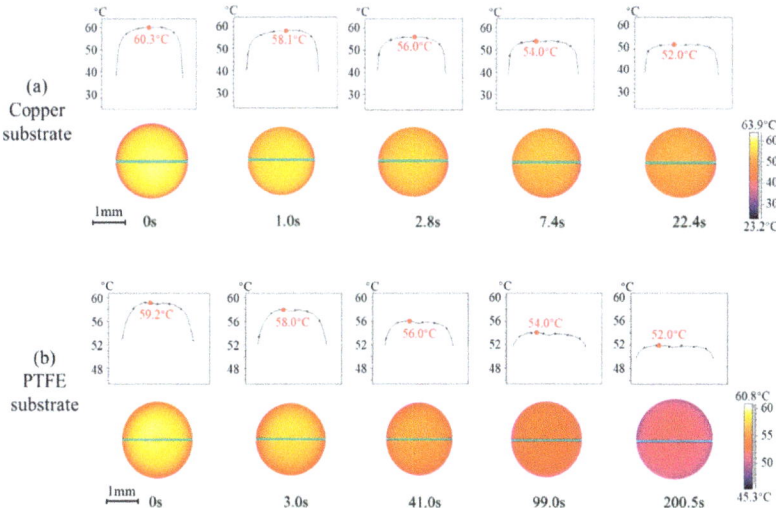

Figure 5. Spatiotemporal evolution of interfacial temperature of LiBr-H_2O droplets on (**a**) copper and (**b**) PTFE substrates coated with FEP.

Figure 6 indicates the representative varying curves of average interfacial temperature of LiBr-H_2O droplets (Figure 6a) and that of pure water droplets (Figure 6b) on a PTFE substrate. For water droplets (Figure 6b), the interfacial temperature is low at the initial period as a result of evaporative cooling, and then increases due to heat supply from the substrate. Moreover, the increase of interfacial temperature speeds up towards the end of the droplet lifetime. This is because, as water evaporates, the volume of water droplet shrinks. In this case, the length for heat conduction from the liquid–solid interface to the liquid–air interface decreases and the heat capacity of the liquid droplet become smaller; therefore, the droplet surface warms up quickly. By comparison, for LiBr-H_2O droplets (Figure 6a), the decrease of interfacial temperature slows down along with time. This is because, as vapor absorption goes on, the volume expansion of the LiBr-H_2O droplet causes an increase in both the characteristic length for heat conduction and the heat capacity of the droplet. In this case, the thermal resistance for heat conduction increases, and the temperature decrease at the droplet surface slows down.

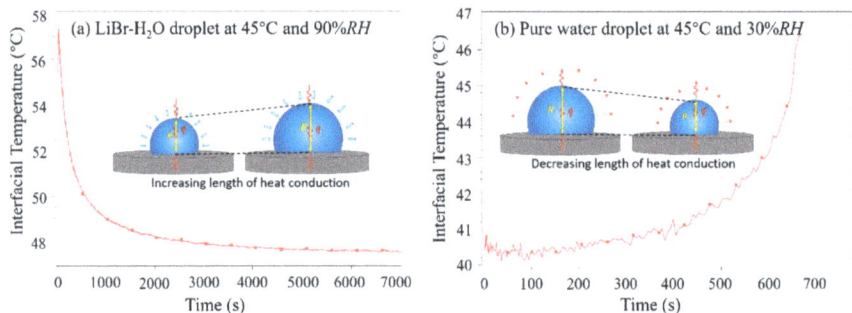

Figure 6. Variation of average interfacial temperature of (**a**) a LiBr-H$_2$O droplets at 45 °C and 90% RH, and (**b**) a pure water droplet at 45 °C and 30% RH on PTFE substrates coated with FEP.

Next, we discuss the evolution of water vapor pressure at the droplet surface. Figure 7 shows the evolution curves as function of interfacial temperature in the case of pure water droplets (saturation line) and LiBr-H$_2$O droplets with different solute concentrations (20 wt.%–60 wt.%), which are calculated according to the *P-T-x* correlations by Patek et al. [21] expressed in Equations (1) and (2). For water droplets, the effect of evaporative cooling will cause a decrease in the interfacial temperature, and therefore a decrease in the saturation vapor pressure at the droplet surface. Taking the condition of 45 °C and 60% RH as an example (for more clear demonstration), the temperature drop induced by evaporative cooling will bring about ~3.0 kPa decrease in the water vapor pressure at the liquid–air interface of pure water droplet, while the overall vapor pressure difference to drive vapor diffusion is only about 4.5 kPa by making use of the isothermal assumption. This indicates that the thermal effect cannot be neglected during droplet evaporation, and the isothermal model is no longer reliable [7,11]. On copper substrates, the high thermal conductivity ensures more efficient heat supply from the substrate so that the interfacial temperature is kept high. In this case, the water vapor pressure at the droplet surface is high, and the evaporation mass flux is large; therefore, the droplet lifetime is shorter than that on a PTFE substrate (Figure 3b).

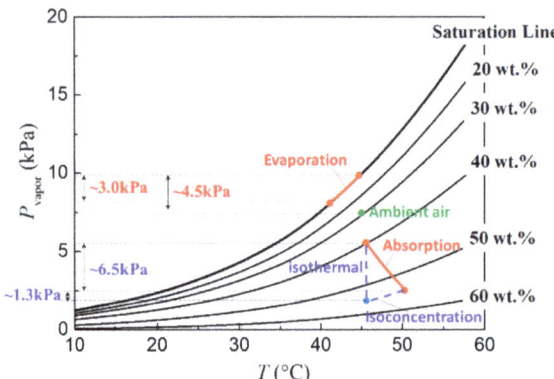

Figure 7. Evolution of water vapor pressure P_{vapor} along with interfacial temperature for pure water droplets during evaporation and LiBr-H$_2$O droplets with different solute concentrations during vapor absorption showed in red lines.

In the case of 54 wt. % LiBr-H$_2$O droplets, the vapor absorption will induce the variation of both temperature and concentration as indicated in Figure 7. By dividing the vapor absorption process into an isoconcentration process (where the temperature changes alone) and an isothermal process (where the solute concentration changes alone), it can be seen that the variation of water vapor pressure induced purely by temperature change is about 1.3 kPa, while the variation induced by concentration change is about 6.5 kPa. This demonstrates that the vapor diffusion in the gas phase and the rate of vapor absorption is mainly controlled by the concentration variations of LiBr–H_2O droplets instead of the interfacial temperature. Since the substrate conductivity only affects the transient heat transfer and the evolution of interfacial temperature, its effect on the rate of vapor absorption is therefore not apparent as evidenced by our experiments. This reminds us that, in the mathematical modeling of vapor absorption, more efforts must be made into the accurate description of the dominating mass transfer process where the concentration distribution is of high importance, while the thermal transport process can be properly simplified in order to improve the simulation efficiency, i.e., computing time.

4. Conclusions

This paper investigates the thermal effects along with vapor absorption into hygroscopic liquid desiccant droplets. The effect of substrate conductivity on the transient heat transfer process and on the rate of vapor absorption is investigated by experiments and theoretical analyses. Results indicate that substrate conductivity plays a crucial role in the transient heat transfer process as demonstrated by the more rapid decrease on the surface temperature of LiBr-H_2O droplets on high-thermal conductivity copper substrates due to efficient heat removal. As a result of the thermal effect, droplets on copper substrates show a slightly higher rate of vapor absorption than those on low-thermal conductivity PTFE substrates. Further analyses by decomposing the variation of water vapor pressure indicate that, compared to the influence of temperature change, the water vapor pressure at the droplet surface is greatly affected by the change of solute concentration during vapor absorption, and therefore the rate of vapor absorption. We conclude that, even though the thermal effect cannot be neglected in the simulation of droplet evaporation as revealed by previous researchers, in the mathematical modelling of the vapor absorption process, the thermal effect can be properly simplified in order to improve the calculation efficiency, and more efforts should be put into accurately capturing the solute diffusion and convection within the droplet.

Author Contributions: Conceptualization, Z.W. and Y.T.; methodology, Z.W. and D.O..; formal analysis, Z.W.; investigation, Z.W.; writing—original draft preparation, Z.W.; writing—review and editing, Z.W. and D.O.; supervision, Y.T. and K.S. All authors have read and agreed to the published version of the manuscript.

Funding: The authors gratefully acknowledge the support received from the International Institute for Carbon-Neutral Energy Research (WPI-I2CNER), and the Japan Society for the Promotion of Science (JSPS). Z.W. acknowledges the support of JST CREST (Grant No. JPMJCR18I1). K.S. acknowledges the support of the European Space Agency (ESA) through grant MAP Project EVAPORATION.

Conflicts of Interest: The authors declare no conflict of interest.

References

1. Chua, K.J.; Chou, S.K.; Yang, W.M. Liquid desiccant materials and dehumidifiers—A review. *Renew. Sustain. Energy Rev.* **2016**, *56*, 179–195.
2. Wang, Z.; Zhang, X.; Li, Z. Evaluation of a flue gas driven open absorption system for heat and water recovery from fossil fuel boilers. *Energy Convers. Manage.* **2016**, *128*, 57–65. [CrossRef]
3. Nath, S.; Bisbano, C.E.; Yue, P.; Boreyko, J.B. Duelling dry zones around hygroscopic droplets. *J. Fluid Mech.* **2018**, *853*, 601–620. [CrossRef]
4. Wang, Z.; Orejon, D.; Sefiane, K.; Takata, Y. Coupled thermal transport and mass diffusion during vapor absorption into hygroscopic liquid desiccant droplets. *Int. J. Heat Mass Transf.* **2019**, *134*, 1014–1023. [CrossRef]
5. Liu, X.H.; Chang, X.M.; Xia, J.J.; Jiang, Y. Performance analysis on the internally cooled dehumidifier using liquid desiccant. *Build. Environ.* **2009**, *44*, 299–308. [CrossRef]
6. Dunn, G.J.; Wilson, S.K.; Duffy, B.R.; David, S.; Sefiane, K. The strong influence of substrate conductivity on droplet evaporation. *J. Fluid Mech.* **2009**, *623*, 329–351. [CrossRef]

7. Dunn, G.J.; Wilson, S.K.; Duffy, B.R.; David, S.; Sefiane, K. A mathematical model for the evaporation of a thin sessile liquid droplet: Comparison between experiment and theory. *Colloids Surf. A Physicochem. Eng. Asp.* **2008**, *323*, 50–55. [CrossRef]
8. Sobac, B.; Brutin, D. Thermal effects of the substrate on water droplet evaporation. *Phys. Rev. E* **2012**, *86*, 021602. [CrossRef]
9. Josyula, T.; Wang, Z.; Askounis, A.; Orejon, D.; Harish, S.; Takata, Y.; Pattamatta, A. Evaporation kinetics of pure water drops: Thermal patterns, Marangoni flow, and interfacial temperature difference. *Phys. Rev. E* **2018**, *98*, 052804. [CrossRef]
10. Talbot, E.L.; Berson, A.; Brown, P.S.; Bain, C.D. Evaporation of picoliter droplets on surfaces with a range of wettabilities and thermal conductivities. *Phys. Rev. E* **2012**, *85*, 061604. [CrossRef]
11. Sefiane, K.; Bennacer, R. An expression for droplet evaporation incorporating thermal effects. *J. Fluid Mech.* **2011**, *667*, 260–271. [CrossRef]
12. Xu, X.; Ma, L. Analysis of the effects of evaporative cooling on the evaporation of liquid droplets using a combined field approach. *Sci. Rep.* **2015**, *5*, 8614. [CrossRef] [PubMed]
13. Zhang, K.; Ma, L.; Xu, X.; Luo, J.; Guo, D. Temperature distribution along the surface of evaporating droplets. *Phys. Rev. E* **2014**, *89*, 032404. [CrossRef] [PubMed]
14. Wang, Y.; Ma, L.; Xu, X.; Luo, J. Combined effects of underlying substrate and evaporative cooling on the evaporation of sessile liquid droplets. *Soft Matter* **2015**, *11*, 5632–5640. [CrossRef] [PubMed]
15. Girard, F.; Antoni, M. Influence of substrate heating on the evaporation dynamics of pinned water droplets. *Langmuir* **2008**, *24*, 11342–11345. [CrossRef] [PubMed]
16. Ristenpart, W.D.; Kim, P.G.; Domingues, C.; Wan, J.; Stone, H.A. Influence of substrate conductivity on circulation reversal in evaporating drops. *Phys. Rev. Lett.* **2007**, *99*, 234502. [CrossRef] [PubMed]
17. Gibbons, M.J.; Di Marco, P.; Robinson, A.J. Heat flux distribution beneath evaporating hydrophilic and superhydrophobic droplets. *Int. J. Heat Mass Transf.* **2020**, *148*, 119093. [CrossRef]
18. Wang, Z.; Orejon, D.; Sefiane, K.; Takata, Y. Water vapor uptake into hygroscopic lithium bromide desiccant droplets: Mechanisms of droplet growth and spreading. *Phys. Chem. Chem. Phys.* **2019**, *21*, 1046–1058. [CrossRef] [PubMed]
19. Orejon, D.; Sefiane, K.; Shanahan, M.E. Stick–slip of evaporating droplets: Substrate hydrophobicity and nanoparticle concentration. *Langmuir* **2011**, *27*, 12834–12843. [CrossRef]
20. Kim, J.H.; Ahn, S.I.; Kim, J.H.; Zin, W.C. Evaporation of water droplets on polymer surfaces. *Langmuir* **2007**, *23*, 6163–6169. [CrossRef]
21. Patek, J.; Klomfar, J. A computationally effective formulation of the thermodynamic properties of LiBr–H_2O solutions from 273 to 500 K over full composition range. *Int. J. Refrig.* **2006**, *29*, 566–578. [CrossRef]

© 2020 by the authors. Licensee MDPI, Basel, Switzerland. This article is an open access article distributed under the terms and conditions of the Creative Commons Attribution (CC BY) license (http://creativecommons.org/licenses/by/4.0/).

Review

Molecule Sensitive Optical Imaging and Monitoring Techniques—A Review of Applications in Micro-Process Engineering

Marcel Nachtmann [†], Julian Deuerling [†] and Matthias Rädle *

Reseach Center CeMOS, Mannheim University of Applied Sciences, Paul-Wittsack-Str. 10, 68163 Mannheim, Germany; m.nachtmann@hs-mannheim.de (M.N.); j.deuerling@hs-mannheim.de (J.D.)
* Correspondence: m.raedle@hs-mannheim.de
† These authors contributed equally to this paper.

Received: 24 February 2020; Accepted: 25 March 2020; Published: 28 March 2020

Abstract: This paper provides an overview of how molecule-sensitive, spatially-resolved technologies can be applied for monitoring and measuring in microchannels. The principles of elastic light scattering, fluorescence, near-infrared, mid-infrared, and Raman imaging, as well as combination techniques, are briefly presented, and their advantages and disadvantages are explained. With optical methods, images can be acquired both scanning and simultaneously as a complete image. Scanning technologies require more acquisition time, and fast moving processes are not easily observable. On the other hand, molecular selectivity is very high, especially in Raman and mid-infrared (MIR) scanning. For near-infrared (NIR) images, the entire measuring range can be simultaneously recorded with indium gallium arsenide (InGaAs) cameras. However, in this wavelength range, water is the dominant molecule, so it is sometimes necessary to use complex learning algorithms that increase the preparation effort before the actual measurement. These technologies excite molecular vibrations in a variety of ways, making these methods suitable for specific products. Besides measurements of the fluid composition, technologies for particle detection are of additional importance. With scattered light techniques and evaluation according to the Mie theory, particles in the range of 0.2–1 µm can be detected, and fast growth processes can be observed. Local multispectral measurements can also be carried out with fiber optic-coupled systems through small probe heads of approximately 1 mm diameter.

Keywords: surface scanning optics; Raman; near infrared; middle infrared imaging; scanning; multimodal spectroscopy; local reaction control techniques; microchannel

---

1. Introduction

In this article, we focus on remote control techniques that may or may not yet be widespread. Most of the presented measurements were carried out at the Center for Mass Spectrometry and Optical Spectroscopy—CeMOS, an interfaculty institution of the University of Applied Sciences in Mannheim, Germany.

Optical time and space resolved measuring technologies in the UV- and visible range are used for the better understanding of flows, mixing processes, and the control of reactions in microchannels. Common image analysis results in two-dimensional images that are measured in reflecting, or if the microchannel has been specially adapted and manufactured, transmitting arrangements. For transmitted light, both the bottom and the lid of the microchannel must be transparent.

Obtaining an increase of contrast is possible in several ways, e.g., by restricting the depth of field from classic image analysis. Only a narrow, defined detection plane results in a sharp image. In transmission, the obtained concentration values are the mean values for the respective vertical axis intercept. Complex tomographic 3D-scanning instruments suppress this effect but require more time to capture the necessarily high number of images.

Classical methods of microscopy and image analysis allow for insights into the spatial distribution of gases and fluids, as well as their temporal development. Contrasting the moving phases greatly improves the optical contrast. Fluorescence marking is common in this context, as it requires low concentrations of markers. The main disadvantage of this technique is the influence of the marker on fluidics and the lack of selectivity. Self-fluorescence detection is only possible in very few cases. These techniques cannot detect the local, time-resolved, concentration-controlled identification of molecular species and a possible tracking of mixtures, inhomogeneity, reactions or deviations in reaction behavior, nor the occurrence of by-products. Therefore, fluid science is in need of measuring systems that combine a high spatial and temporal resolution with molecular selectivity and no disturbance of fluidics. For optical technologies, the measurement range is, depending on the detector used, not limited by the range of perception of the human eye. All known approaches have advantages and disadvantages, and the current state-of-the-art approaches are explained in following article.

2. Materials and Methods

In this paper, an overview of how optical measurement techniques can be used to quantify effects in microchannels is given. To illustrate this, the interaction of light with a specific target or matter in general is explained. Light, in this case, extends from the ultraviolet range via the visually visible range via near-infrared to mid-infrared and should not—as in common speech—be limited to the visible range only.

Light interacts with the product, both elastically and inelastically. Elastic light scattering means that the incoming photon does not cause a molecular change of the energy states and leaves the medium with the same wavelength. Inelastic light scattering means that vibrations or rotations of the molecules, as well as the lattice vibration phonons of the solid bodies, are excited [1,2]. Electronically excited states can also be achieved, in particular by irradiating with ultraviolet light [3].

First, it is necessary to note which basic beam paths geometries, like transmission, backscattering, or remission into other angles, can be applied for the measurement techniques to be described. Figure 1 headlights the different types of possible interactions between light and a particle.

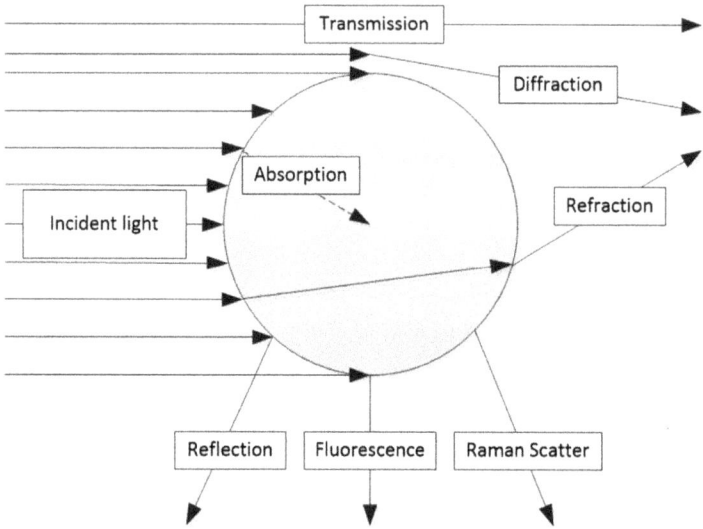

Figure 1. Types of interaction between light and a particle [4].

Based on the different interaction types, a variety of measurement techniques are possible. One of the most commonly used techniques is the transmission setup. A change of the outgoing light in transmission can take place through the absorption of certain wavelengths by particles, other disperse substances such as drops, aerosols, or through the molecular absorption of fluids or gases [2,5].

In fluids, the dissolved substances are often referred to as molecularly-disperse in order to distinguish them from macroscopically disperse substances like particles [6]. For solutions, i.e., molecularly-disperse dissolved substances, transmission is one of the most frequently used application areas [3]. Absorbed light can be described with the Lambert–Beer law according to the following formula in which E stands for extinction or absorption, I_0 is the emitted, I is the transmitted light, ε is the extinction coefficient, c is molar concentration, and d is the layer thickness: [3]

$$E_\lambda = \log_{10}(I_0/I) = \varepsilon \cdot c \cdot d \quad (1)$$

The decisive factors here are the molar concentration c and the extinction coefficient ε of the irradiated wavelength. Usually, the experimental setup defines the layer thickness d. For disperse materials, leaving the primary optical axis for measuring the absorption in combination with scattering is possible.

Backscattering, side scattering, and forward scattering are common angular possibilities [3]. The basic theory for disperse materials is accumulated in almost all literature in the application of the Mie scattered light theory and simplifications of the range of small particles or large particles. In the range of small particles, the Rayleigh approximation is often used, as it is applicable when the light is smaller than about a quarter of the irradiated wavelength. For large particles, Fraunhofer diffraction is commonly used. Here, the particles have to be larger than about five times the wavelength of the exciting light. All these theories imply that the particles are round spheres. Theories for non-round particles are almost unheard of [7–12].

By using transmission, the extinction coefficient can be determined from the intensity loss, which includes the phenomena of absorption and scattering [3]. Applied to the geometries of microchannels, a measurement setup based on transmission has the decisive disadvantage of the need to be able to illuminate through the microchannel. This requires two transparent side surfaces. Some applications may need temperature control. With this setup, temperature control is usually more problematic than using non-transparent metallic boundaries. In transparent designs, the heat transfer coefficient is a further limitation. Therefore, the pure transmission arrangement is usually not suitable for microchannel applications.

Another geometry is the so-called attenuated total reflection technique (ATR). A light beam, irradiated laterally at a certain angle, is reflected at the boundary between the crystal or optical material of the sensor and the fluid. With the corresponding theory, the quantum-mechanically explainable evanescent wave penetrates a few wavelengths into the medium and is potentially absorbed there. The ATR technique is relevant when high concentrations and high extinction coefficients simultaneously occur. This effect can be found in the ultraviolet range below the wavelength of 320 nm, i.e., UV-B or UV-C systems. With dyes, the effect also take place in the visible range. Mostly, the ATR setup is used in the mid-infrared range, because the excitation of the ground states of vibration at this point is accompanied by high extinction coefficients [3].

Light scattering measurements are used for dispersed-phased products or if a change in the disperse phase is a suitable control or quality parameter for such processes. Precipitation reactions, for instance, with particles precipitating from a molecularly-dissolved starting material that reaches supersaturation or grows in a particular way such that scattering is increased, are also suitable for scattering measurements. Optical scattering correlates in the same way as transmission via the Mie scattering theory or via Rayleigh or Fraunhofer diffraction by using the same mathematical equations. The scattered light measurements can be separated into geometrical subdomains. A decisive technological question and boundary condition is the arrangement regarding the angle of the scattered light. In a laboratory environment, 90° scattering is common, but this scattering angle is difficult to adapt when applied to microchannels [7–10]. Target process variables are mostly local concentrations. In microchannels, for example, local particle concentrations and the velocities of particle size and concentration changes are relevant variables. The determination of kinetic parameters using these process variables is possible.

2.1. Particle Detection

The relevant particle size for many products is about one micron. For theoretical description, the Mie theory, postulated by Gustav Mie, suits best. Currently, several programs are available for the calculation of diffraction patterns; these include the free-to-use algorithm by Wiscombe et al. [12]. An example for such diffraction patterns is the angle dependence of scatter light intensities. Mostly, a logarithm scale best suits the matching of a wide illumination distribution for different angles. This requires special detectors to suit the whole intensity distribution. For different particle sizes, the angle dependence, which is a special boundary condition in Mie theory, differs. Mie back scattering is shown for 180° in Figure 2 [7,10].

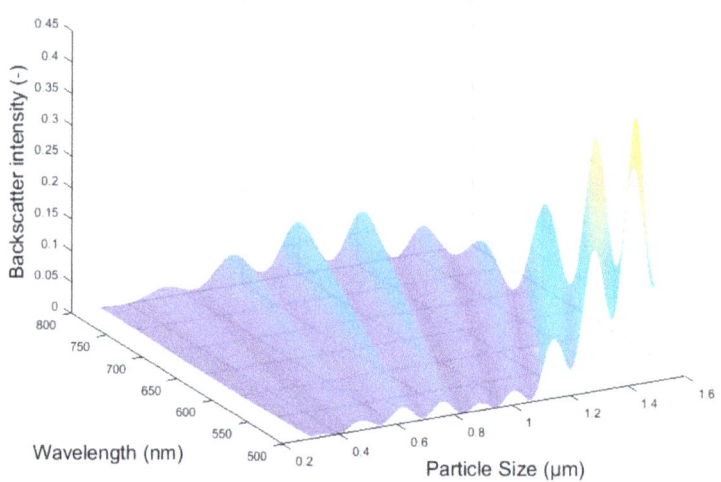

Figure 2. Mie back scattering intensity at 180° vs. particle size and wavelength [7].

In addition, the quotient of the wavelength and particle size correlate with angle dependence. This means that the particle growth can be determined by using several different wavelengths in the visible area. Usually, two wavelengths are sufficient. Different wavelength bands can be obtained by using different light-emitting diodes (LEDs) instead of a halogen lamp or by using different optical filters in front of the detector [11]. Broadband detectors accumulate all irradiated light. By using different LEDs, simultaneous measurements are, out of the box, not possible. Optical filters are needed or the LEDs have to be used in an alternating fashion. The oscillating intensity quotient, obtained from those two wavelengths, correlates with the actual particle size and matches the theoretical calculations following Mie, as shown in Figures 3 and 4.

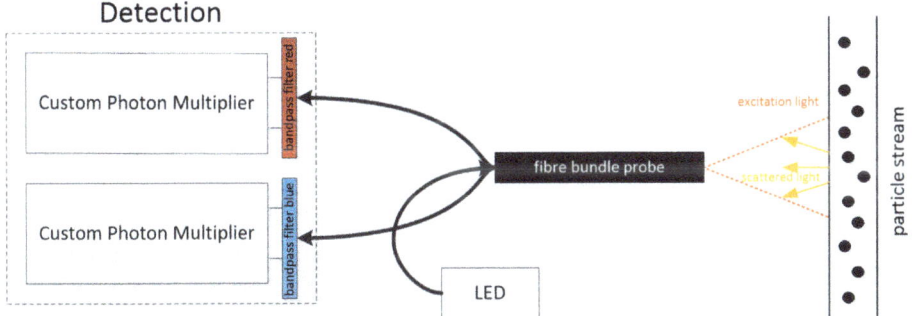

Figure 3. Detection setup with different bandpass filters [4].

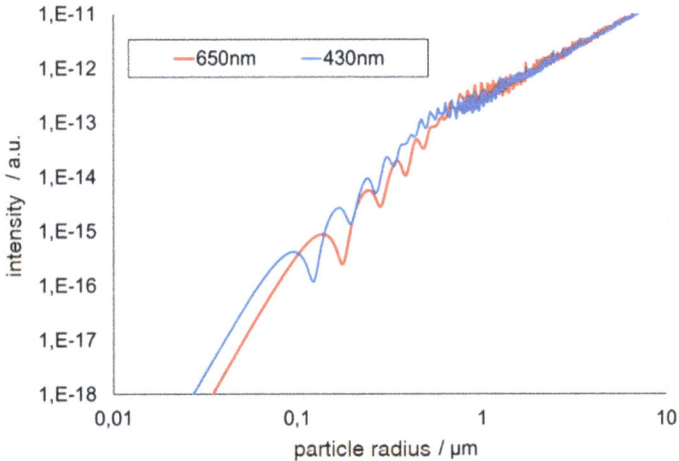

Figure 4. Oscillating intensity from two wavelengths [4].

Several limitations may apply, but particle size can, in most cases, be determined with an acquisition rate of over 1000 Hz. Limitations may be the particle size itself. Only in the Mie area is the quotient oscillating. If particles are smaller 100 nm, this effect cannot be observed. The presented technique is also not applicable for particles in the Rayleigh area. Growing monodisperse particles can be measured up to about 2 µm. For bigger particles, the oscillation is too frequent to be separated for each particle. Mistakes and mix-ups presumably increase with ascending particle size. This method may be limited to a specific particle size range, but it performs well with small component size and the measurement frequency [8,10,11]. Furthermore, probes for scattered-light measurements can be extremely small. For example, manufacturing backscatter probes even in a blunt cannula with an inner diameter of 1.1 mm is possible [13].

In such a blunt cannula (Figure 5), several glass fibers can be included for different light sources and detectors. This technology is used for measurements in micro channels to obtain particle size, concentration, and growth [13].

The method is certainly special and is limited to the specified particle size range, but, compared to all other measuring methods, it has the huge advantage of being small, i.e., miniaturized, and of high-speed detection. For the questions posed here, scattering at an angle of 180°, i.e., backscattering, is often a more suitable arrangement. Additionally, 45° backscattering— or 135° from the point of view of angle geometry—is a popular arrangement. The 135° scattering angle has particularly established itself in the field of color pigment monitoring [8,10,11]. The entire optical arrangement of the emitter and the detector can be realized in one stable, steady instrument as an easy-to-use setup with a small physical gap between the transmitter and the receiver. This leads to reduced surface contamination effects in the measurement signal—bypassing the Tyndall effect (Figure 6).

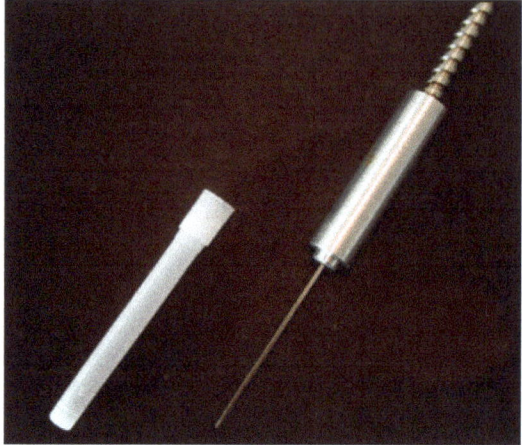

Figure 5. Cannula sterican 19G, 30° bevel [4].

Figure 6. Tyndall effect, 10 ppm fluorescein sodium in water [4].

In addition to elastic light scattering, spectroscopic methods like molecular excitations are relevant, especially for the quantification of molecule concentrations. Here, the whole range of spectral technologies is available and is explained in the following sections with regard to their theoretical explanations and these technologies' possible applications.

2.2. Ultraviolet/Visible (UV/VIS) Spectroscopy

Starting with UV spectroscopy, the electrons are usually excited into higher orbitals by so-called electronic excitations. These are usually characterized by high extinction coefficients and have very wide bands or absorption edges. Therefore, UV spectroscopy has a high sensitivity but a low selectivity. Real process control is only possible in exceptional cases where the molecules make this possible [3,14]. Examples can be found in the conversion of nitrobenzene and sulfuric acid to nitrobenzene sulfonic acid. In this process, the nitrobenzene band disappears, and, thus, distinguishing nitrobenzene concentrations from the beginning to the end of the process is easily possible. In VIS spectroscopy, the same effects occur as in UV spectroscopy, though only for substances that are usually recognizable as colors in the visible spectrum. Therefore, the effects occur with dyes (molecularly-disperse dissolved dyes) or with color pigments (particulate substances containing chromophores) [4].

2.3. NIR-Spectroscopy

Near-infrared (NIR) spectroscopy uses the fact that almost all fluids in the NIR range provide absorption bands. Thus, they are usually distinguishable. In the world of chemical monitoring, these methods are highly established because it is easily possible to guide the light via glass fibers to chemical reactors [3,14,15]. Miniaturized fiber optic probes are also suitable for microchannels [4]. The excitation of the second or higher harmonics of the vibrational–rotational transitions of molecules caused by the photons is used. The fundamental oscillation of the same molecules is in the mid-infrared range [3,14,15]. The extinction is relatively weak and only measurable at higher concentrations (above approx. 0.1% in fluids) [4].

The photons used in near infrared spectroscopy usually excite the harmonics of present molecules. Furthermore, excitations can occur through combination oscillations while supplying the necessary energies. The incoming photon must therefore trigger an OH oscillation in the water molecule and additionally excite an oscillation in the bending angle of the water molecule via the so-called banding mode. The water molecule, in particular, is very dominant in the near infrared range. The coupling of the electromagnetic wave of the photon is a dipole interaction, so molecules with a strong dipole moment have high extinction coefficients in the near infrared range [16].

Due to the large number of possible combinations of oscillations and rotations, the near-infrared spectra are complex. For the pictorial measurements discussed in this article, two possible methods are convenient: a scanning method, which records a complete spectrum per point, or photometric methods, where only one wavelength is recorded at a time but can be displayed over a large area by using infrared cameras. Both methods have advantages and disadvantages. An advantage of the spectroscopic method is that a wavelength resolution of the laboratory apparatuses that measure the samples. The whole range of known reaction monitoring is available. A disadvantage is the time required for the measurement. About 1 s is required per measuring point, so a flat image takes several hours to measure. An alternative are area measurements. Broadband photo filters or light-emitting diodes in the near-infrared range on the transmitter side are necessary. The advantage here is that an image is quickly acquired (about 40 ms), but only one wavelength is recorded. The interpretability of the results is therefore often limited because the installed filters or light sources do not get close to reaching the wavelength resolution of the spectrometers. For this type of application, we later focus on mixing processes.

2.4. Mid-Infrared Spectroscopy

The mid-infrared region is characterized by the excitation of the ground states of vibration; the theory is sufficiently well-known in the literature. The significant advantage of mid-infrared spectroscopy is its selectivity in combination with its high sensitivity with regard to molecules and their detection limits on the concentration scale [3,16].

With this type of spectroscopy, fundamental oscillations of the molecules are scanned with, in contrast to NIR spectroscopy, much higher excitation cross sections. The advantage of this is the possibility of detecting even small concentrations down to the parts per million range. Due to the sharper bands, the selectivity towards NIR is also greatly increased. A disadvantage is the complex technology required, which means that a robust design of the device is difficult to realize. Existing fiber technologies are unstable, have limited spectral range, and have no long-term stability in harsh environments. The extremely high absorption of water also proves to be disadvantageous. Water-containing substance systems are superimposed by water absorbance, and other substances—especially with low concentrations—are only found with difficulty [3,16].

2.5. Fluorescence Spectroscopy

In fluorescence spectroscopy, the incident light excites short-lived electronic states within the molecule. After typical times in the nanosecond range, the molecules return to their ground state. Often, however, they do not fall back to the same oscillation state of the electronic ground state, instead falling to excited oscillation levels. For this reason, the wavelength emitted is often longer, and the photon energy is thus lower than that of the incident light. The fluorescence is extremely sensitive to detection. However, it requires a molecule-specific electronic state. Therefore, only a part of the molecule fluoresces, and imaging is limited with regard to the selection of molecules [4,14,17].

2.6. Raman Spectroscopy

Raman technology is the strongest upcoming technique at the moment. It scans the fundamental oscillation of molecules, but, in contrast to mid-infrared (MIR) spectroscopy, it does so by exploiting the polarizability of molecules. Stimulated with photons in the visible range, the molecules are short-time excited into a virtual state and fall back instantaneously into another vibrational–rotational state of the electronic basic level. This has the advantages that, depending on the used excitation wavelength, classical fiber-optic sensors are applicable for Raman measurement techniques and the high selectivity of MIR spectroscopy is present. On the other hand, the Raman effect is extremely weak. Strong lasers used as excitation sources and long integration times are necessary. Technological advancements in the last few years have had a strong and positive impact on this situation [3,15,18–20].

Table 1 shows a comparison of all mentioned spectroscopic methods for better comparison, highlighting the major advantages and disadvantages of each technique.

Table 1. Summary of the different measurement techniques.

Technique	Pro	Contra
Ultraviolet/Visible (UV/VIS) Spectroscopy	high sensitivity	low selectivity
NIR (Near-Infrared) Spectroscopy	easily accessible	weak extinction; only measurable at higher concentrations
NIR (Scanning)	high range of detection and monitoring	slow
NIR (Photometric)	fast	smaller wavelength resolution
Mid-Infrared Spectroscopy	high selectivity in combination with a high sensitivity	difficulties with water due to superimposition
Fluorescence Spectroscopy	extremely sensitive	imaging is limited with regard to the selection of the molecule; invasive
Raman Spectroscopy	high selectivity	very weak effect; strong laser or very sensitive detector needed
Particle Detection	very fast	only works for specific particle size

3. Results

In this next part, a different application, based on the results, is presented.

3.1. NIR Image Analysis

Analogous to image analysis in the UV/VIS range, charge-coupled device (CCD) cameras are used, and wavelength selectivity with optical filters is utilized either on the emitter or on the detection side. Cameras on silicon-based chips have a limited spectral range and are not sensitive in the NIR wavelength region. Instead, cameras based on indium gallium arsenide (InGaAs) are used, and these are sensitive in the 900–1600 nm range. Exact wavelength ranges may vary depending on the exact InGaAs detector used. In combination with NIR LEDs, acquiring images with wavelength-selected brightness levels according to the absorption bands of fluids for locally-fluctuating concentrations is possible. Acquiring reference images without the presence of any absorbing substances is important. The computer-based optimization of images, like anti-shading methods, can be used to compensate for inhomogeneous illumination problems. Advantage of NIR LED illumination as an alternative to broadband illumination and the use of sequentially used filters include the simplicity of its design and its fast change of wavelengths. Without special equipment, image sequences of 25 full frames/s are achievable, and, thus, moderately fast changing processes are accessible for detection. With individual wavelengths, pulses down to microseconds can also be captured [21]. By mathematically linking the remission at the surface with the absorption of the investigated fluids, extinction data and the resulting layer thickness distributions can be visualized. In Figure 7, you can see the measurement data of water strands.

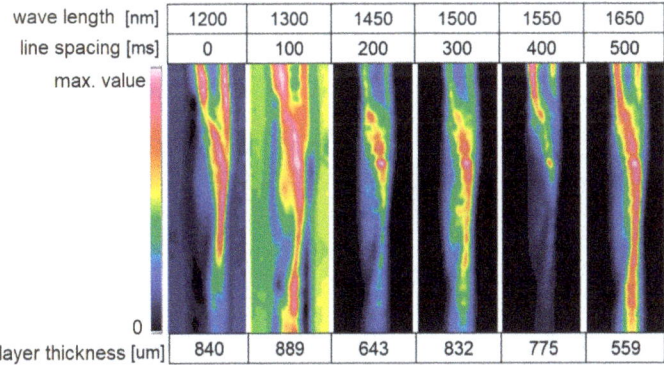

Figure 7. Near-infrared (NIR) measurement on water strands at all different wavelengths 274 × 142 mm (220 × 220 DPI), water glycerin [21].

A uniform illumination of the examined surface is important for later measurements, especially with curved surfaces (e.g., pipes). Here, the refractive index difference to the gas space causes light deflection, which can lead to the significant misinterpretation of the measurement signals. In the present case, the deflections were already significantly suppressed. This reduction is possible by distributing the light directions via a special dome illumination. The work so far has concentrated on the application of film thickness measurements. In addition, reactive processes, where both the fluid concentration and the film thickness distribution changes, are to be measured [21].

3.2. MIR Image Analysis

In the field of thermal imaging, mid-infrared (MIR) image analysis is common. However, depth of field and contrast are usually not sufficient for scientific purposes. The method of MIR scanning has proven to be more favorable. Commercial devices typically reach scanning velocities of one-to-two measuring points per second. Therefore, the acquisition of images is very time-consuming. A two-stage procedure for data acquisition proves to be more capable. The first stage, image scanning with full spectrum width, is selected for each pixel. Here, the assignment of interesting statements to the suitable wavelengths can be assigned via gold standards for narrowly-defined local conditions [22]. In the following step, the interesting wavelengths and target cutout are selected and monitored with up to 300,000 measurements per second with a spatial resolution of 20 μm in the fast scanning mode using quantum cascade lasers for routine sample measurements [23].

The special confocal beam path provides a tolerance towards distance variations to the target. As proven, a fast, flat, and confocal absorption measurement in the middle infrared range, from 3 to 5 µm, is possible (Figure 8). With these techniques, unequal surfaces with topographies can be sampled, and coatings with a layer thickness of less than one micrometer are eligible for molecule selective detection. It is possible to transfer the measuring principle to further applications [23].

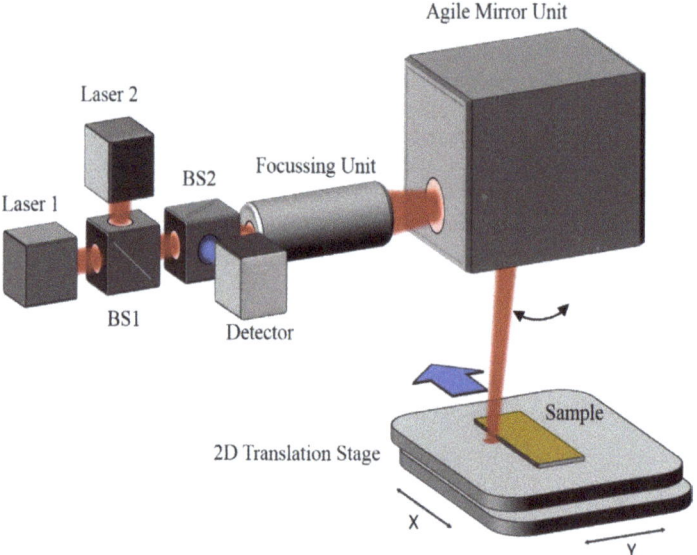

Figure 8. Schematic view of the mid-infrared (MIR) scanner setup with confocal mounted detector [23].

3.3. Raman Scanning Image Analysis

As explained above, Raman spectroscopy requires high laser power and long exposure times. With precisely focused and installed Raman probe-heads (Figure 9) 2D Raman scanning is possible [24]. Even 3D Raman scanning seems to be achievable.

Figure 9. Raman scanning probe with long working distance mounted on a 3D-displaceable table [24].

In the measurements conducted so far, this method has made it possible to visualize 2D Raman scans in fluids for the molecular quantification of the concentration profiles in microchannels and to perform the fundamental investigation of the mixing processes of different accessible fluids, as shown in Figure 10 [24].

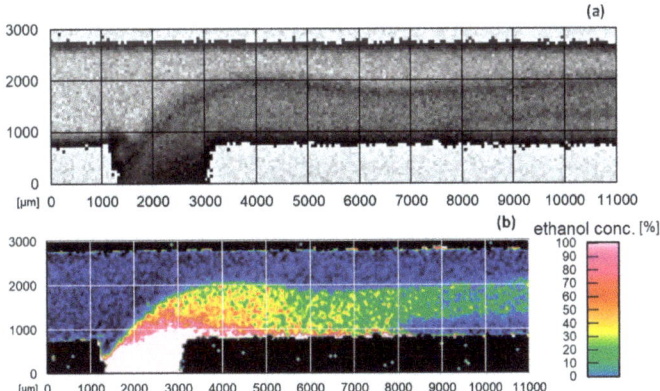

Figure 10. (a) 11 × 3 mm Raman scan of ethanol in the reactor and (b) 11 × 3 mm false-color image of the ethanol concentration curve in the reactor calculated from quotient method [24].

The measurement rate correlates with the spectral resolution. Increasing the measurement rate decreases the possible resolution. Further advancement towards higher measurement speeds is possible by limiting the numbers of detected Raman shifts by replacing classical Raman spectroscopy by Raman multichannel photometry [18,24]. Large-area detectors promote Raman photometry by replacing the necessary dispersion and thus small-area detections in the spectrometer with filter techniques and single photon counters (such as a photomultiplier [25]). The area and sensitivity gain lead to a considerable increase in measurement speed. Spectral information is lost, but this can be of minor importance for known materials [18,26].

Multichannel-Raman photometry shines for process control with rather simple matrices. In Figure 11, the successful reaction tracing of binding CO_2 to an amine until saturation is plotted, with a comparison of spectroscopic and photometric measurements [18]. The absorption of CO_2 is well-known in the literature [27]. In photometers, faster detection rates, due to single photon counting detectors, are available and are subsequently able to resolve the reaction progress in a more detailed fashion. Both the spectrometer and photometer are able to trace the reaction progress. Complex matrices demand an increasing number of measurement channels. The measurement speed remains, but at least the financial advantages fade over the number of channels implemented compared to a spectrometer system. The sweet spot for Raman photometry, combining fast measurements and cost reduction, is between two and four channels. Raman photometers use an excitation laser as a light source. Every measurement channel is equipped with at least one small-band optical filter and one single-photon counter. A beam-splitter has to be added for a two-channel-photometer, and three beam-splitters for a 4-channel-photometer if an equal signal distribution is favored [18].

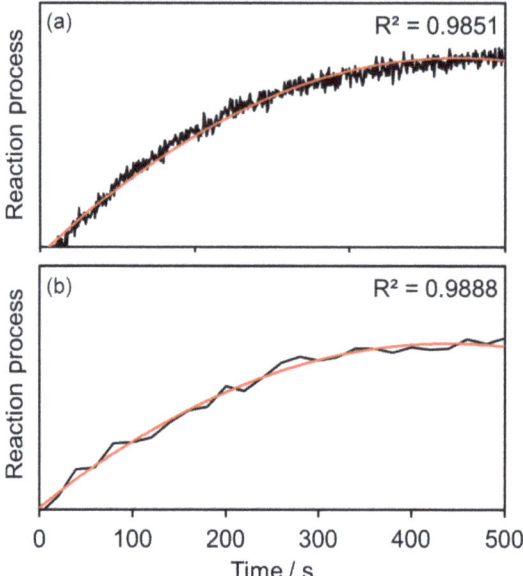

Figure 11. Comparison of (**a**) photometric and (**b**) spectroscopic measurements of CO_2 binding to amine [18].

3.4. Scattering Light Measurement Technology

Scattering light techniques (elastic light scattering) can be used to control the formation of disperse phases, e.g., due to product failures. Fiber optic backscattering was used with a detection rate of several full spectra per second [9]. The fast detection rate is mandatory to visualize the precipitation of inorganic products in the millisecond range. Microchannels, in particular, allow for strong concentration gradients and, thus, pH or temperature gradients. This can lead to extremely fast precipitation processes. Fast measurement technologies are required. In geometries with constant flow and events occurring constantly at each location, kinetic fields for precipitation processes can be developed in combination with fast backscatter measuring systems with scanning equipment [4].

3.5. Fluorescence Techniques

A method for malignant tissue detection is a laser-induced fluorescence, noncontact-imaging approach, as shown in Figure 12.

The malignant tissue is labeled with a fluorescent marker. The selected marker can be excited in the first optical window of the tissue. The optical window is defined as a wavelength region between the absorption band of hemoglobin (Hb) and water. The emission also has to be in the first optical window to avoid absorption. This method is not supposed to replace common technologies, like computer tomography (CT) or magnetic resonance imaging (MRI), but it is meant to supplement the highlighting of the malignant areas to reduce time that is needed with common technologies. The advantages of this method, compared to common technologies, are its greatly increased measurement rate and low price [28].

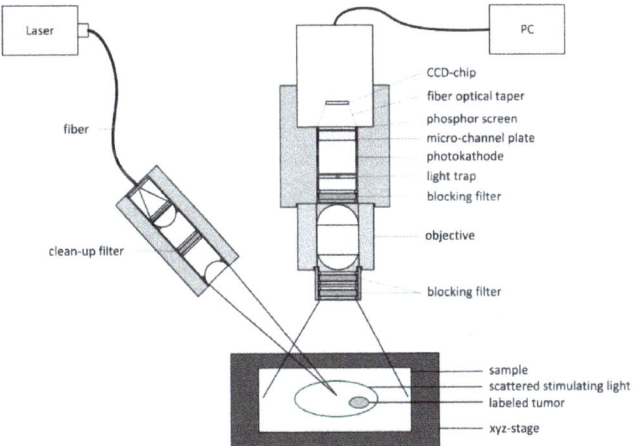

Figure 12. Schematic diagram of a fluorescence imaging system for malignant tissue detection [28].

3.6. Combination Techniques: UV/VIS/NIR/Fluorescence

This example illustrates how the different modes described can be built into one device. This allows for, e.g., the distribution of lipids via Raman scanning. The additional information of the local moisture that is gathered from NIR scanning is helpful for gathering information on wound healing and also for reaction monitoring in the microchannel. An optimal control technology seems to be the combination of different techniques and the selective wavelength information generated by different wavelength regimes. The combination provides a flexible application for divers scientific problems.

The backscattering needle probe was designed as a setup for simultaneously, multimodal spectroscopic measurements. The basic setup, presented in Figure 13, contains glass fibers for ultraviolet, visible, near infrared and fluorescence spectroscopy.

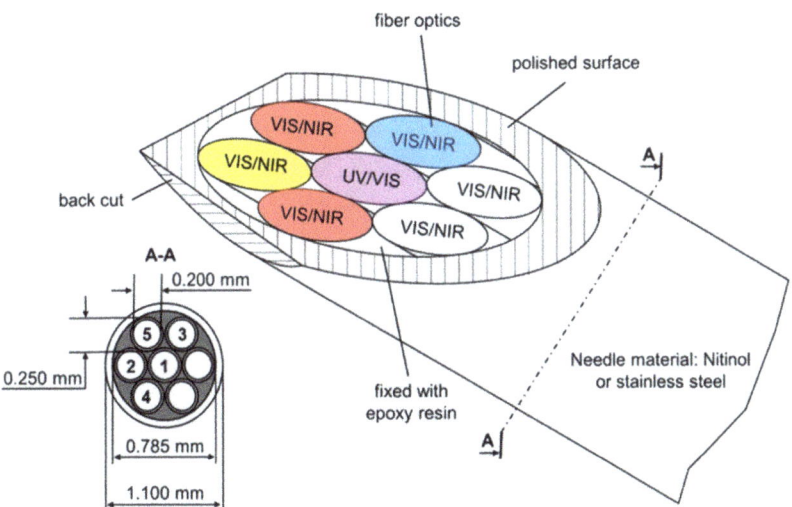

Figure 13. Needle probe for multispectral backscattering measurements. Ultraviolet (UV), visible (VIS), NIR, and fluorescence measurements are possible. The needle is optimized for tissue penetration with a front and back cut [13].

The probe head consists of a name-giving needle and seven bare glass fibers for illumination and detection [13]. Glass fibers can be used to transport light from the probe head to the spectrometer or detection unit [29,30]. The real time detection of malignant tissue in in vivo measurements is possible. A back cut can be applied for better tissue penetration. Measurements can be done with either a spectrometric or a photometric set-up. For evaluation purposes, the measurements can be done with a spectrometer, and a virtual photometer can be calculated. This approach has the advantages that it has been used for proof of concept photometric measurements and reference measurements can be simultaneously obtained. In addition, the received data are compressed. The monitoring of several metabolic parameters, like Hb, deoxy Hb, scattering, fat, and auto fluorescence, is possible. Clustered in groups for malignant, marginal, and healthy tissue, the obtained data are analyzed by a 2D principle component analysis (PCA). All samples can be assigned to their respective groups. Classical histology validates all measurements [13].

4. Discussion

In summary, the processes in microchannels, focused on the local concentration distribution of molecules, can be isolated and processed when the target molecules are spectroscopically accessible. In currently available technology, this usually means that molecules can be excited to vibrate. This applies, for example, to water, hydrocarbons, nitrates, and phosphates but not to the dissolved portion of hydrochloric acid, hydrogen fluoride, hydrogen bromide, or other ion concentrations without covalent bonds.

In principle, two classes of devices can be distinguished with regard to pictorial representations:

1. Scanning systems.
2. Simultaneous imaging systems.

Scanning systems are generally slower. Therefore, current technological development is focused on reducing the necessary number of wavelengths and increasing their detection sensitivity. These two measures promote an increase in throughput. For known molecules and non-complex matrices, the spectral resolution of spectrometers can be traded for faster measurement rates in photometers.

The current level is approx. 300,000 measuring points/ss for MIR, NIR, and UV/VIS, and it is possibly 1000 measuring points/ss in relation to Raman measurement technology [23].

The differences between the scanning technologies result from their respective applications and boundary conditions. In the presence of water, the strong water absorption in the MIR and NIR often covers the more interesting spectra of the searched molecules that may be present in smaller concentrations. Here, Raman technology proves to be beneficial, since the dipole character of the H_2O molecule has limited relevance. On the other hand, Raman spectra show low extinctions, excluding thin layers and very low concentrations.

Molecule selectivity increases from shorter to longer wavelengths: UV < VIS < NIR < MIR. However, with longer wavelengths and the associated higher molecular selectivity, technological effort also increases. The selectivity of Raman spectroscopy is similar to that of MIR measurements: comparatively high.

With simultaneous imaging flat camera systems, all pixels are simultaneously recorded during a measurement. The advantage of simultaneity is often accompanied by a loss of wavelength selectivity and, thus, molecular selectivity. Because only one wavelength can be recorded at a time, the filter or selective illumination is, because of technical limitations, always somehow broadband. In the case of the necessary planar illumination, compromises usually have to be made with regard to detection optics, which leads to a loss of image quality. Contrast reduction, shading, and image distortion are the visible effects of this process.

Ultimately, the decision for a suitable measuring system is made based on the application and the resulting boundary conditions.

5. Patents

Patent Nr. DE102018105067A1: Bildgebendes System zur nichtinvasiven optischen Untersuchung von Gewebe in der Tiefe; Ahlers, Rolf, Prof. Dr., 64625, Bensheim, DE; Braun, Frank, 69226, Nußloch, DE; Hien, Andreas, 68161, Mannheim, DE; Rädle, Matthias, Prof. Dr., 67273, Weisenheim am Berg, DE.

Patent Nr. DE102014107342A1: Vorrichtung und Verfahren zur Erkennung von Krebstumoren und anderen Gewebeveränderungen; Braun, Frank, 69226, Nußloch, DE; Gretz, Norbert, Prof. Dr., 68259, Mannheim, DE; Rädle, Matthias, Prof. Dr., 67273, Weisenheim am Berg, DE.

Author Contributions: All authors contributed equal to the wording of each section. All authors have read and agreed to the published version of the manuscript.

Funding: German Federal Ministry of Education and Research, grant number 03FH8I01IA, funded this research.

Acknowledgments: In this article, the authors draw on contributions from many members of the CeMOS Research Center. In particular, we would like to thank Tim Kümmel, Isabel Medina, Annabell Heintz, Lukas Schmitt, Tobias Teumer, Stefan Schorz, Steffen Manser and Julia Siber for providing technology, data and images.

Conflicts of Interest: The authors declare no conflict of interest. The funders had no role in the design of the study, in the collection, analyses, or interpretation of data, in the writing of the manuscript, or in the decision to publish the results.

References

1. der Optik, L. Lichtstreuung. Available online: https://www.spektrum.de/lexikon/optik/lichtstreuung/1859 (accessed on 21 February 2020).
2. Ritgen, U. *Analytische Chemie I*; Springer: Berlin/Heidelberg, Germany, 2019. [CrossRef]
3. Kessler, R. *Prozessanalytik*; Wiley-VCH: Weinheim, Germany, 2006; ISBN 978-3-527-31196-5.
4. CeMOS–Centre for Mass Spectrometry and Optical Spectroscopy, 68163 Mannheim, Germany
5. Redmond, H.E.; Dial, K.D.; Thompson, J.E. Light Scattering and Absorption by Wind Blown Dust: Theory, Measurment, and Recent Data. *Aeolian Res.* **2009**. [CrossRef]
6. Kopeliovich, D. Classification of Dispersions. Available online: https://www.substech.com/dokuwiki/doku.php?id=classification_of_dispersions (accessed on 23 February 2020).
7. Ross-Jones, J.; Teumer, T.; Capitain, C.; Tippkötter, N.; Krause, M.; Methner, F.-J.; Rädle, M. Analytical Methods for In-line Characterization of Beer Haze. In Proceedings of the Trends in Brewing, Ghent, Belgium, 8–12 April 2018. [CrossRef]
8. Ross-Jones, J.; Teumer, T.; Garcia, F.; Krause, N.; Rädle, M.; Methner, F.-J.; Nirschl, H. Particle Size Characterization of Precipitated Non-Spherical Protein Particles. In Proceedings of the EBC 2017, European Brewery Convention, Ljubljana, Slovenia, 14–18 May 2017.
9. Teumer, T.; Capitain, C.; Ross-Jones, J.; Tippkötter, N.; Rädle, M.; Methner, F.-J. In-line Haze Monitoring Using a Spectrally Resolved Back Scattering Sensor. *Brew. Sci.* **2018**, 49–55. [CrossRef]
10. Teumer, T.; Rädle, M.; Methner, F. Possibility of monitoring beer haze with static light scattering, a theoretical background. *Brew. Sci.* **2019**, *72*, 132–140. [CrossRef]
11. Teumer, T.; Ross-Jones, J.; Schlachter, K.; Lerche, D.; Methner, F.-J.; Rädle, M. Stoffspezifische optische Streuwirkung bei Rekristallisation und Fällung. In Proceedings of the Jahrestreffen der Fachgruppe LVT, Leipzig, Germany, 4–6. March 2015.
12. Wiscombe, W. Improved Mie Scattering Algorithms. *Appl. Opt.* **1980**, *19*, 1505–1509. [CrossRef] [PubMed]
13. Braun, F.; Schalk, R.; Nachtmann, M.; Hien, A.; Frank, R.; Beuermann, T.; Methner, F.-J.; Kränzlin, B.; Rädle, M.; Gretz, N. A customized multispectral needle probe combined with a virtual photometric setup for in vivo detection of Lewis lung carcinoma in an animal model. *IOP Sci.* **2019**. [CrossRef]
14. Hesse, M.; Meier, H.; Zeeh, B.; Bienz, S.; Bigler, L.; Fox, T. *Spektroskopische Methoden in der organischen Chemie*, 9th ed.; Georg Thieme Verlag: Stuttgart, Germany; New York, NY, USA, 2016.
15. Schmidt, W. *Optical Spectroscopy in Chemistry and Life Sciences: An Introduction*; Wiley-VCH: Weinheim, Germany, 2005.
16. Günzler, H.; Gremlich, H.-U. *IR-Spektroskopie: Eine Einführung*; Wiley VCH Verlag: Weinheim, Germany, 2003; ISBN 978-3-527-66285-2.
17. Lakowicz, J. *Principles of Fluorescence Spectroscopy*; Springer: Boston, MA, USA, 2006; ISBN 978-0-387-46312-4.

18. Nachtmann, M.; Keck, S.; Braun, F.; Eckhardt, H.S.; Mattolat, C.; Gretz, N.; Scholl, S.; Rädle, M. A customized stand-alone photometric Raman sensor applicable in explosive atmospheres: A proof-of-concept study. *J. Sens. Sens. Syst.* **2018**. [CrossRef]
19. Pudlas, M. *Nicht Invasive Diagnostik in der Regenerativen Medizin Mittels Raman-Spektroskopie*; Fraunhofer Verlag: Stuttgart, Germany, 2012; ISBN 978-3-8396-0396-3.
20. Weidlein, J.; Müller, U.; Dehnicke, K. *Schwingungsspektroskopie: Eine Einführung*; Thieme: Stuttgart, Germany; New York, NY, USA, 1982.
21. Medina, I.; Schmitt, L.; Kapoustina, V.; Rädle, M.; Scholl, S. Untersuchung von lokalen Schichtdickenverteilungen in Fluiden mithilfe der Nahinfrarot-Bildanalyse. *Chemie Ingenieur Technik* **2019**. [CrossRef]
22. Rabe, J.-H.; Sammour, D.A.; Schulz, S.; Munteanu, B.; Ott, M.; Ochs, K.; Hohenberger, P.; Marx, A.; Platten, M.; Opitz, C.; et al. Fourier Transform Infrared Microscopy Enables Guidance of Automated Mass Spectrometry Imaging to Predefined Tissue Morphologies. *Sci. Rep.* **2018**. [CrossRef]
23. Kümmel, T.; Teumer, T.; Dörnhofer, P.; Manser, S.; Heinrich, S.; Hien, A.; Marx, J.; Methner, F.-J.; Rädle, M.; Wängler, B. Absorption Properties of Lipid-Based Substances by Non-Invasive fast Mid-Infrared Imaging. In *OCM 2019-Optical Characterization of Materials: Conference Proceedings*; KIT Scientific Publishing: Karlsruhe, Germany, 2019; pp. 23–33.
24. Deuerling, J.; Manser, S.; Siber, J.; Keck, S.; Hufnagel, T.; Schmitt, L.; Medina, I.; Repke, J.-U.; Rädle, M. Schnelle lokale Raman-Messung zur Untersuchung von Konzentrationsprofilen bei Mischvorgängen in Mikrokanälen. *Chem. Ing. Tech.* **2019**, *83*, 593. [CrossRef]
25. Demtröder, W. *Laserspektroskopie: Grundlagen und Techniken*; Springer: Berlin/Heidelberg, Germany; New York, NY, USA, 2007.
26. Braun, F.; Schwolow, S.; Seltenreich, J.; Kochmann, N.; Röder, T.; Gretz, N.; Rädle, M. Highly Sensitive Raman Spectroscopy with Low Laser Power for Fast In-Line Reaction and Multiphase Flow Monitoring. *Anal. Chem.* **2016**. [CrossRef] [PubMed]
27. Vogt, M.; Pasel, C.; Bathen, D. Characterisation of CO 2 absorption in various solvents or PCC applications by Raman Spectroscopy. *Energy Procedia* **2011**, *4*, 1520–1525. [CrossRef]
28. Hien, A.; Pretze, M.; Braun, F.; Schäfer, E.; Kümmel, T.; Roscher, M.; Schock-Kusch, D.; Waldeck, J.; Wängler, C.; Rädle, M.; et al. Noncontact recognition of fluorescently labeled objects in deep tissue via a novel optical light beam arrangement. *PLoS ONE* **2018**, *13*, e0208236. [CrossRef] [PubMed]
29. Schröder, G.; Treiber, H. *Technische Optik: Grundlagen und Anwendungen*, 11th ed.; Vogel: Würzburg, Germany, 2014.
30. Hering, E.; Bressler, K.; Gutekunst, J. *Elektronik für Ingenieure und Naturwissenschaftler*, 6th ed.; Springer: Berlin/Heidelberg, Germany; New York, NY, USA, 2014.

© 2020 by the authors. Licensee MDPI, Basel, Switzerland. This article is an open access article distributed under the terms and conditions of the Creative Commons Attribution (CC BY) license (http://creativecommons.org/licenses/by/4.0/).

Article

Velocity Measurements in Channel Gas Flows in the Slip Regime by means of Molecular Tagging Velocimetry

Dominique Fratantonio [1,*], Marcos Rojas-Cárdenas [2], Christine Barrot [2], Lucien Baldas [2] and Stéphane Colin [2]

[1] Los Alamos National Laboratories, Physics Division, Los Alamos, NM 87545, USA
[2] Institut Clément Ader (ICA), Université de Toulouse CNRS, INSA, ISAE-SUPAERO, Mines-Albi, UPS, 31400 Toulouse, France; marcos.rojas@insa-toulouse.fr (M.R.-C.); christine.barrot@insa-toulouse.fr (C.B.); lucien.baldas@insa-toulouse.fr (L.B.); stephane.colin@insa-toulouse.fr (S.C.)
* Correspondence: dfratantonio@lanl.gov; Tel.: +1-505-665-6572

Received: 5 March 2020; Accepted: 30 March 2020; Published: 2 April 2020

Abstract: Direct measurements of the slip velocity in rarefied gas flows produced by local thermodynamic non-equilibrium at the wall represent crucial information for the validation of existing theoretical and numerical models. In this work, molecular tagging velocimetry (MTV) by direct phosphorescence is applied to argon and helium flows at low pressures in a 1-mm deep channel. MTV has provided accurate measurements of the molecular displacement of the gas at average pressures of the order of 1 kPa. To the best of our knowledge, this work reports the very first flow visualizations of a gas in a confined domain and in the slip flow regime, with Knudsen numbers up to 0.014. MTV is cross-validated with mass flowrate measurements by the constant volume technique. The two diagnostic methods are applied simultaneously, and the measurements in terms of average velocity at the test section are in good agreement. Moreover, preliminary results of the slip velocity at the wall are computed from the MTV data by means of a reconstruction method.

Keywords: rarefied gas; slip velocity; channel flow; molecular tagging velocimetry

1. Introduction

In the last twenty years, microelectromechanical systems (MEMS) have received a lot of attention due to their appealing properties of low volume, low weight, low energy consumption, and system integrability. The development of new microfabrication techniques has made the design of novel MEMS possible, highly increasing the range of possible practical applications. There are many examples of MEMS that found a commercial application and that are present in many of our daily life utilities, such as microaccelerometers for smartphones, micronozzles [1] for space applications, microactuators [2] for aeronautical applications, micro heat exchangers [3], and Knudsen pumps [4], to name just a few.

Thus, from a scientific point of view, several new interesting physical problematics have arisen. Microsystems behave quite differently from their corresponding macro versions. As the device becomes smaller, the inertial forces decrease, and surface forces tend to gain importance with respect to volume forces. In gas microflows, the dynamics at the molecular level can lead to local non-equilibrium that affects the macroscopic thermodynamic properties of the gas flow. This microeffect is called rarefaction, and it is a consequence of the relatively low number of intermolecular collisions inside the control volume. The Knudsen number, $Kn = \lambda/L_c$, where λ is the molecular mean free path and L_c is the characteristic size of the system, identifies the degree of gas rarefaction. Most of the microfluidic devices operate in slightly rarefied gas conditions, with Kn in the range $(10^{-3}; 10^{-1})$, which corresponds to the slip flow regime. This regime can be attained either in microfluidic devices with a low characteristic length L_c or in larger systems at low pressure. In this rarefaction regime, the classical continuum representation of the flow is still acceptable for fluid particles that are far enough from the system's solid boundaries, i.e., at a distance larger than the well-known Knudsen layer thickness, which is considered to be of the same order of magnitude of λ. For rarefied fluid systems that involve wall–gas interactions, the number of collisions between the gas molecules and the wall surfaces gains importance with respect to intermolecular collisions, thus producing a thermodynamic non-equilibrium state of the gas in the vicinity of the wall. At a macroscopic level, this produces a velocity slip and a temperature jump at the wall [5], i.e., non-negligible discontinuities of the kinematic and thermodynamic properties between surface and fluid. These rarefaction effects strongly influence the heat and mass transfers. Therefore, a correct modeling of the thermodynamic disequilibrium at the wall is crucial for an accurate prediction of the mass flow rates and heat fluxes within the microdevice. Experimental observations and analysis of the rarefaction effects are needed for the validation and the uncertainty quantification of the numerous theoretical and numerical analyses that already exist. However, most of the experimental analyses available in the literature on rarefied gas flows in channels are carried out by measuring global quantities, such as the mass flow rate, inlet and outlet pressures, and temperatures, to indirectly quantify the effects given by the velocity slip and temperature jump at the wall [6,7].

To the best of our knowledge, there are no experimental data in the literature that provide local information on the velocity field in confined rarefied gas flows. In this context, we have developed an optical velocimetry technique with the final aim of applying it to rarefied gas flows inside a channel. Molecular tagging velocimetry (MTV) is an optochemical, low-intrusive technique capable of providing local measurements of the velocity field in gas flows. The technique is based on the tracking of the displacement of a suitable molecular tracer that exhibits photoluminescence properties. While various versions of this technique currently exist [8,9], this work demonstrates the implementation of 1D-MTV by direct phosphorescence, which is based on the long-lived phosphorescence emission of specific molecules, which can be activated by a single UV laser line excitation. In this work, the molecular tracer used is acetone (CH_3COCH_3) vapor, which has a pressure vapor of 24 kPa at 20 °C and is recognized to have low toxicity if inhaled. A pulsed laser source tags a region of interest inside a gas flow and a photodetector captures the light emitted by the tracer at two successive times. The basic functioning principle of 1D-MTV in a plane channel is schematized in Figure 1. At the reference time t_0 a pulsed laser beam tags the initial reference position of the tracer molecules along a line perpendicular to the main flow direction. As the tracer moves by following the flow streamlines, the initial tagged line starts to deform accordingly to the velocity profile of the gas. The photodetector captures an image of the reference position at t_0 and an image of the deformed tagged line at a time $t_0 + \Delta t$, where Δt is called the delay. The velocity profile can then be inferred from the molecular displacement profile $s_x(y)$ of the tagged line.

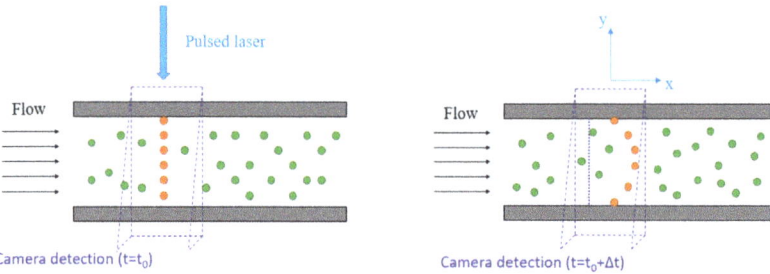

Figure 1. Basic principle of 1D-molecular tagging velocimetry (MTV) by direct phosphorescence, for a gas flowing in a plane channel from left to right.

MTV is often considered as the molecular counterpart of particle image velocimetry (PIV). The substantial difference between the two velocimetry techniques relies on the nature of the tracer. MTV uses molecular seeding, which does not involve the complications specific to particle-based techniques, such as the relatively high response time of the particles, the difficulties in generating submicrometric sized particles, and the inaccuracies introduced by the Brownian motion noise. These issues make PIV inaccurate for applications in confined rarefied gas flows. For this reason, MTV is currently the most promising technique for flow visualization in microdevices and for accurate measurements of slip velocities at the wall. Nevertheless, while the MTV technique has already been successfully applied in various gas flows systems [10,11], the application of MTV to the case of confined and rarefied gas flows has been prevented up to now by various technological challenges. MTV measurements of fluid microflows require the adoption of short laser pulses at relatively high energy with a beam diameter smaller than the characteristic length of the system. Since diffraction phenomena limit the minimum laser beam waist that can be physically obtained by means of focusing optical lenses to about 30–150 μm, the channel height cannot be smaller than about 1 mm in order to be able to accurately resolve the strong velocity gradients. Hence, in order to reach Knudsen numbers in the slip flow regime, the average pressure needs to be lowered down to about 1 kPa. Unfortunately, low average pressures reduce the amount of tracer molecules present in the gas flow and increase the molecular diffusion of the tracer through the background gas. The combination of these two low-pressure effects makes the light signal disappear much faster than in high-pressure conditions, thus making MTV applications in rarefied gas flows very challenging.

In previous works we demonstrated that MTV could provide good results in a millimetric rectangular channel for a non-rarefied gas flow at atmospheric pressure and ambient temperature [12]. However, a combined effect of advection and molecular diffusion of the tracer in the background gas flow produces an important molecular displacement at the wall that is entirely independent of the actual slip velocity of the gas. Due to this phenomenon, known as Taylor dispersion, it is not possible to deduce correct values of the velocity profile by simply deriving the displacement profile with respect to time, not even in gas flows at atmospheric pressure. Subsequently, we have proposed a numerical method based on the advection–diffusion equation that was able to correctly reconstruct the velocity profile from the displacement of the tagged line. In a first step, the reliability of the method was proven by applying it to direct simulation Monte Carlo (DSMC) numerical experiments [13]. Later on, the method was proven to be valid also in the case of physical MTV experiments in confined non-rarefied flows. The experiments were performed in a millimetric channel at atmospheric and subatmospheric pressures down to a minimum average pressure of 42 kPa [14]. At this pressure level in a 1-mm deep channel, the flow was still in the continuum regime. In order to obtain velocity profile measurements in the slip flow regime, lower ranges of pressure are necessary. However, phosphorescence lifetime is highly reduced at such low pressures, and in order for MTV to work the tagged regions need to be interrogated on a time interval that is smaller than the lifetime of the tracer emission. In this respect, a very important step forward was realized by experimentally characterizing the photoluminescence properties of the tracer molecules in terms of signal intensity and lifetime at low pressures. The experiments performed on acetone and diacetyl vapors demonstrated that MTV measurements in confined rarefied gas flows should be possible [15].

Thus, by using all the knowledge acquired up to now on this subject, we present in this work optical experimental results of molecular displacement in the slip flow regime in a confined environment. To the best of our knowledge, the results here presented are the very first flow visualizations of gas displacement in a confined domain and in rarefied conditions. Moreover, they are the first visual proof of velocity slip at the wall in rarefied gas flows. MTV data are reported for argon and helium flows in a 1-mm deep channel with rectangular cross-section and at pressures between 30 and 0.9 kPa, thus reaching Knudsen number values up to $Kn = 0.015$. In this work, the Knudsen number Kn is defined with respect to the channel height H.

This manuscript is organized as follows. Section 2 describes in detail the experimental setup used for carrying out MTV measurements in low-pressure gas flows through a channel with a rectangular cross-section. In Section 3, the experimental methodology used for acquiring MTV images in slow-varying thermodynamic gas flow conditions is thoroughly explained and the advantages and disadvantages of this approach are critically discussed. Section 4 describes the sequence of used post-processing operations, from the recorded MTV images to the extraction of the displacement profile data to be used in the reconstruction method developed in [13]. In Section 5, the experimental results of this work are reported. Firstly, MTV data of argon and helium flows in non-rarefied and rarefied conditions are presented. Secondly, MTV data are compared with data obtained by the constant volume (CV) technique in terms of average velocity. Finally, the conclusions and future developments of this work are discussed in Section 6.

2. Experimental Setup

The gas system designed and built in this work for applying MTV to gas flows at low pressures is illustrated in Figure 2. The basic working principle of this open loop gas system relies on (i) preparing the gas-tracer mixture in a big reservoir at the desired pressure p_1 and tracer concentration χ and (ii) forcing the gas mixture to pass through the channel by pumping it into the atmosphere. The upstream reservoir was composed of two tanks of about 90 L each. A pipe with a large diameter of about 1.5 cm and a length of 1.3 m connected the two tanks. At the outlet of the channel, two GEFI® rotary vacuum pumps (Pfeiffer Vacuum, Aßlar, Germany) of the same model were used in parallel to achieve the highest mass flow rate through the channel. The pumping at the channel outlet was required for exhausting the gas from the tanks into the atmosphere since the experimental flow conditions were at pressures lower than the atmospheric one. The gas system was equipped with Inficon Sky® CDG025D capacitance diaphragm gauges (Inficon, Bad Ragaz, Switzerland) for measuring the gas pressure at different locations. Pressure sensors P1 and P2 were used for monitoring the static pressure at the inlet and the outlet of the channel, respectively. Pressure sensor P3 on the mixing line was necessary for monitoring the complete outgassing of air in the acetone bottle. Pressure sensors P4 and P1 could be both used for measuring the pressure in the tanks. P4 could also be used during the experiments in order to verify that the pressure inside the tanks is at any time equal to the one at the entrance of the channel. For all the experiments carried out during this work, no difference was recorded between the measured p_1 and p_4 values.

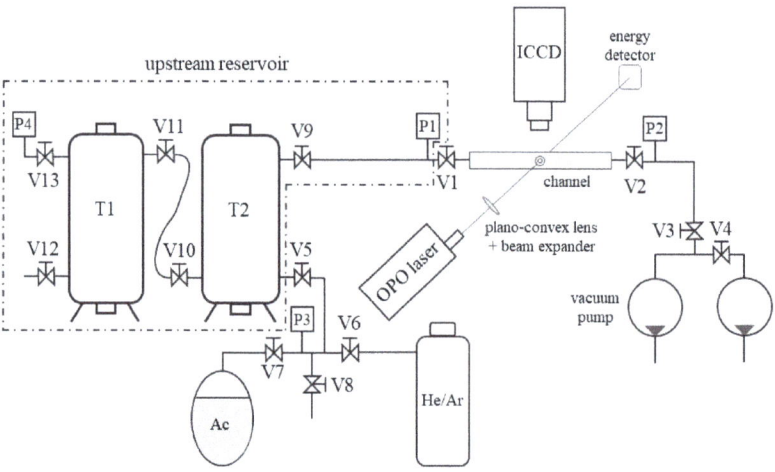

Figure 2. Gas system for application of MTV to the gas–vapor mixture flows at low pressures. T1 and T2 are two tanks; P are pressure sensors; V1 is the inlet channel valve; and V2 is the outlet channel valve.

The choice of using an open loop gas circuit instead of an apparently simpler closed loop system is the result of several practical issues that a closed loop configuration has for the application of MTV.

Firstly, it is difficult to find commercial pumps or fans that are simultaneously oil-free, leakage-free, and chemically resistant to acetone vapor, and, secondly, the tested channel itself is not perfectly free from air leakages. Since oxygen molecules are efficient quenchers of acetone phosphorescence [16,17], a fresh gas–acetone mixture needs to be introduced in the test section at each MTV acquisition, thus making the closed loop configuration an unsuitable solution. The main disadvantage in using an open loop gas system is that the thermodynamic flow conditions in the tested channel vary in time. The inlet–outlet pressure difference and the average pressure at the test section decreased in time as the upstream reservoir pressure decreased. This consideration justifies the use of a couple of big tanks for the upstream reservoir to minimize the variations in time of the flow conditions in the test section. The stability of the experimental conditions in relation to the MTV acquisitions was discussed in detail in Section 3. Nevertheless, a varying upstream gas pressure turned out to be an advantage as it allowed the simultaneous application of the CV technique for monitoring the mass flow rate through the channel. This simultaneous measurement technique provided further experimental data that was used for MTV data validation.

The channel was manufactured with (i) an optical access for the UV laser excitation of the tracer seeded in the gas flow; (ii) an optical access for capturing the phosphorescence emission of the tracer by means of an intensified CCD; (iii) materials that were chemically resistant to acetone vapor; and (iv) good sealing properties for preventing leakages at low pressures. Two PEEK® plates (Victrex plc, Lancashire, UK) composed the main body of the channel. The channel was grooved inside the first plate, while the second one served as a cover. A planar joint of Kalrez® (DuPont de Nemours, Inc., Wilmington, DE, USA) was positioned in between these two plates in order to prevent leakages.

Let us define a Cartesian coordinate system in which the cross-section of the channel lies on the y-z plane, and the x-coordinate represents the position along the channel length L. The inlet and the outlet of the channel corresponded, respectively, to $x = 0$ and $x = L$, while the walls were located at $y = \pm H/2$ and $z = \pm b/2$, where H and b are the height and the width of the channel, respectively. The laser beam excitation was centered at $x = L/2$ and $z = 0$ (Figure 3b). As shown in Figure 3a, an optical access in borosilicate, transparent to light in the visible spectrum, allowed the acquisition of the phosphorescence signal in the x-y plane. Figure 4a reports an image of the channel recorded by the CCD through the borosilicate glass. Two other windows were used for the laser beam access. They were fabricated in a special grade of Suprasil® (Heraeus, Hanau, Germany), which is a high purity synthetic fused silica material highly transparent to UV light. The first window allowed the laser beam access to the channel section by one side and the second window allowed its exit by the other side. This double-window system prevented laser reflections in the interrogation region and allowed a continuous monitoring of the laser energy at the laser exit by means of an energy detector. Figure 4b illustrates a sketch of the structural composition of the channel cross-section in the y-z plane and at $x = L/2$. As it can be observed, the value of the channel height H at the position of the laser beam excitation directly depended on the actual distance between the two Suprasil® windows.

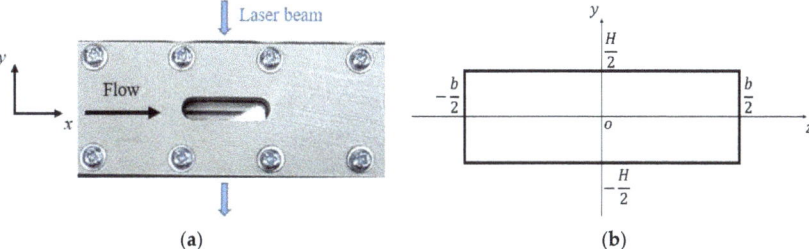

Figure 3. In (**a**), a global view of the test section along the channel from the point of the CCD camera. The blue arrows indicate the optical access for the laser beam and the black arrow shows the gas flow direction. In (**b**), a sketch of the channel cross-section along with the adopted Cartesian coordinate system.

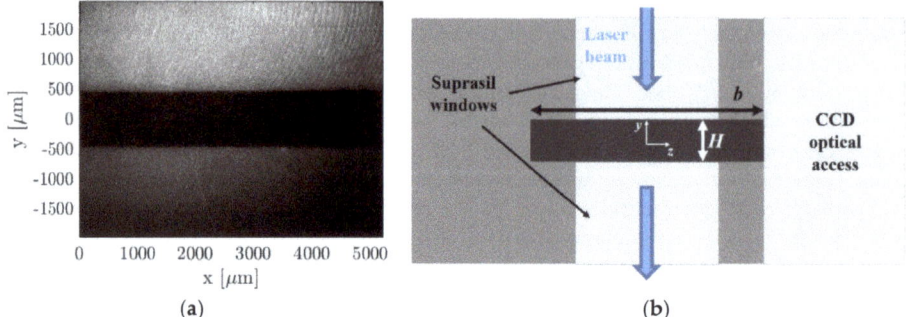

Figure 4. In (**a**), an image of the channel gap in the region of interest for MTV, recorded with the CCD. In (**b**), a sketch of the channel cross-section at the MTV test section. The upper and lower bright regions correspond to the Suprasil® windows for the laser beam access. The bright region on the right represents the optical access for the CCD camera. The width of the cross-section and the distance between the Suprasil® windows are indicated on the image as b and H, respectively.

Accurate measurements of the channel height H at the MTV test section are required for accurately extracting the local velocity profile from the MTV displacement data by means of the reconstruction method. For the computation of the average velocity from the mass flow rate measurements provided by the CV technique (described in Section 5), reliable measurements of both width b and height H at the MTV test section are necessary. With the help of tomographic imaging, the position of the Suprasil® windows has been carefully adjusted to be in line with the walls of the channel grooved into the PEEK® plate. Tomographic images also showed proper conditions of the PEEK® and the windows surfaces. However, because of the low average pressure in the gas system used during MTV acquisitions, the windows position can vary slightly as a function of the pressure difference between the inside and the outside of the channel. Tomographic measurements could be made only with the channel at atmospheric pressure. The uncontrolled micrometric movements of the windows are then sources of uncertainty for the measured value of H. For this reason, an alternative solution has been considered for having the best estimation of the wall distance H. This approach was presented in detail in Section 4 and consisted of an image processing procedure that detects the real positions of the wall directly on MTV images. By this methodology, the channel height was measured to be $H = 1003$ μm, which was consistent with the value measured from tomographic images. The width b was measured to be 5000.8 μm with a standard deviation of 7.4 μm by tomographic measurements, and the length L of the channel was 20 cm.

For the molecular tagging system, an OPOlette HE355LD laser was used for all the experiments, since it allowed us to provide the proper wavelength excitation for acetone vapor, which was set to 310 nm. The laser pulses were shot at a rate of 20 Hz and last 7 ns each. The diameter of the beam was focused into the interrogation area down to about 150 μm. The beam diameter was here defined as the full width at half-maximum, i.e., the width at which the laser energy was half of the central peak value. After UV excitation, acetone vapor emitted a certain amount of light that could be described as the sum of two components: (i) a strong light emission that lasts for only some nanoseconds, which is commonly defined as fluorescence, and (ii) a less intense light emission that can last up to a millisecond, which is known as phosphorescence. The definitions of fluorescence and phosphorescence emissions are not based on the intensity and the duration of the emission but on the type of intramolecular electronic transition that makes an excited molecule come back to its ground electronic state [18]. An extensive analysis of the dependency of acetone phosphorescence intensity and lifetime on excitation wavelength and pressure is discussed in Fratantonio et al.'s research [15]. The laser beam diameter has been estimated from images of the fluorescence intensity distribution.

A 12-bit Imager Intense (LaVision®) progressive scan CCD coupled with a 25-mm intensified relay optics (IRO) was employed for the image recording. The IRO was made of a S20 type photocathode and a P46 phosphor plate. The CCD was composed of 1376 photodiodes × 1040 photodiodes. Due to the very low intensity of the phosphorescence emission, a binning process is required for increasing the sensitivity to light of the CCD at the expense of reducing the spatial resolution. A 4 × 4 binning was used for all the experimental data presented in this work; the resolution of the CCD was thus reduced with a total of 344 pixels × 260 pixels, each pixel corresponding to 16 photodiodes. A 105 mm f: 2.8 and an inverted 28 mm f: 2.8 Micro Nikkor lenses (Nikon Inc., Tokyo, Japan) were used for collecting the light emitted by the acetone molecules. The external optical system and the internal optical collector of the IRO and CCD provided an overall magnification of about 1.7. As the CCD covered an actual area of 8.87 mm × 6.71 mm, the field of view was 5.29 mm × 4 mm. Consequently, each pixel corresponded to an area of 3.8 µm × 3.8 µm without binning and to 15.2 µm × 15.2 µm with a 4 × 4 binning. A programmable timing unit (PTU) was used for synchronizing the laser trigger, the camera shutter, and the IRO trigger. The main parameters that can be controlled for each record are the IRO gate Δt_{gate}, i.e., the time interval of light integration, the IRO amplification gain G, the delay Δt between the IRO trigger and the laser excitation, and the exposition time t_{CCD} of the CCD detector.

The low phosphorescence emission rate and the low amount of tracer molecules seeded in a low-pressure gas flow made the amount of light emitted after one single laser excitation very low. In addition, the laser energy level needs to be limited to avoid damaging the Suprasil® windows. If the amount of light collected from one laser excitation is too low to stand out from the CCD background noise, averaging more images does not help in increasing the signal intensity. For this reason, the on-chip integration was used, which allowed us to collect in one single image the light generated by more than one laser excitation. The final image is the result of averaging N_i images, each one integrating N_l laser pulses. A higher number N_i of averaged images can increase the quality of the resulting image by reducing the data fluctuations, and thus increases the signal-to-noise ratio (SNR). More information on the signal acquisition can be found in Fratantonio et al.'s research [15]. The uncertainty on the velocity measurement depends on the accuracy of the measured molecular displacement and the time separation between two images.

3. Experimental Methodology

In order to be able to measure a sufficiently deformed tagged line by means of MTV, stationary and relatively high flow rates are needed. The desired velocity magnitudes through the test section were obtained by imposing large pressure differences between the inlet and the outlet of the channel. Nevertheless, the MTV experiments were performed in time-dependent experimental conditions, since the imposed pressure difference was not constant with time. However, as it will be demonstrated in this section, the flow conditions were stable enough during the time needed to perform one MTV acquisition.

With reference to the sketch of Figure 2, the preparation of an experimental run conducted at low pressure was done by filling the upstream reservoir (tanks T1 plus T2) with a gas mixture at a pressure p_1 of about 5 kPa and by vacuuming to a minimal pressure the volume downstream from valve V1, which was initially closed. When opening valve V1 at the channel inlet, the gas mixture started flowing through the channel and the downstream pressure p_2 quickly increased to about 1.2 kPa, as a consequence of the limited maximum mass flow rate that the two pumps could sustain at the working pressure p_2. Figure 5 shows the typical evolution in time of the pressure conditions applied to the tested channel during one experimental run at low pressure.

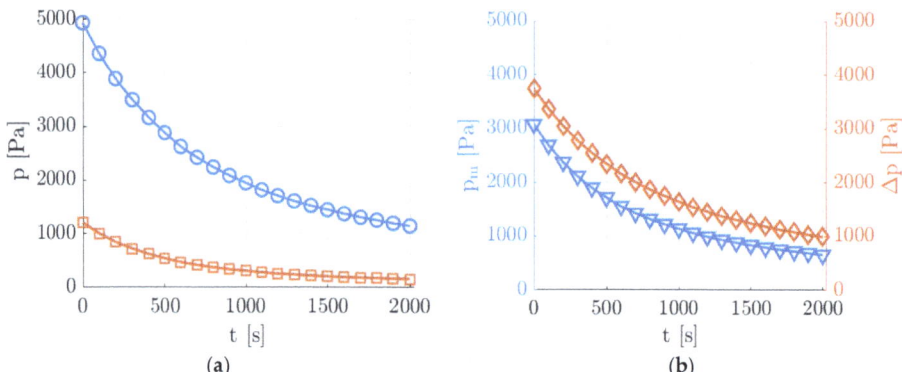

Figure 5. Evolution in time of the experimental pressure conditions generated at low pressures acquired by pressure sensors P1 and P2: (**a**) upstream p_1 (○) and downstream p_2 (□) pressure measurements and (**b**) mean pressure $p_m = \frac{p_1+p_2}{2}$ (▽) and pressure difference $\Delta p = p_1 - p_2$ (◇).

The average pressure p_m in the channel varied between 3000 and 700 Pa with a corresponding pressure difference Δp that decreased from 4000 to 1000 Pa. For a helium–acetone flow with acetone molar concentration of $\chi = 20\%$, this pressure range corresponded to a value of Kn at the test section varying from 0.004 to 0.018. The decrease of p_m and Δp slowed down as the average pressure in the upstream reservoir decreased since the mass flow rate through the channel decreased as well. The slowly decreasing downstream pressure was beneficial for the stability of the velocity profile to be measured. The magnitude of the velocity profile was strongly dependent on the pressure gradient, and the fact that the upstream and downstream pressures decreased simultaneously made the pressure gradient vary slower. Moreover, as the pressure in the system decreased, the pressure in the upstream tanks varied slower as well because of the decreasing mass flow rate through the tested channel. At the lowest average pressure, the flow speed was thus much more stable, which is a suitable experimental condition for the application of MTV. Even though the experimental run could continue towards even lower pressures than those reported in Figure 5, this work presents MTV results only down to a minimum pressure of about 900 Pa.

The application of MTV for the investigation of the velocity field requires the gas flow to be steady or at least slowly varying in time. The evolution in time of the investigated velocity profile during one experimental run could be predicted and its rate of variation could be compared with the time requirements of the acquisition system. The recording of one image with $N_l = 100$ and with a laser repetition rate of 20 Hz took 5 s. From the pressure measurements, it was possible to calculate that the velocity profile varied by less than 0.5% during the recording of one single image at any time along the experimental run. This fact confirms that the experimental setup was able to generate experimental flow conditions that were in the slip regime and that were stable enough for MTV acquisitions. Nevertheless, the SNR of one single recorded image was too low to provide an accurate measurement of the tagged line displacement. Figure 6 shows an example of one single image collecting 100 laser pulses with an intensifier IRO gate $\Delta t_{gate} = 100$ ns per laser shot.

Exploitable results for carrying out the velocity measurement require averaging several images representing the same flow conditions. A possible strategy consists of averaging consecutive images recorded along one experimental run by taking advantage of the slow variation in time of the investigated flow conditions. Since one complete experimental run lasted about 30 min and the laser repetition rate had a frequency of 20 Hz, a group of about 360 images could be collected for $N_l = 100$ laser shots per image. While for a time frame duration of one image the thermodynamic conditions were substantially the same, different images recorded along one experimental run were related to different average pressures and flow rate conditions. Figure 7 illustrates the averaging procedure used.

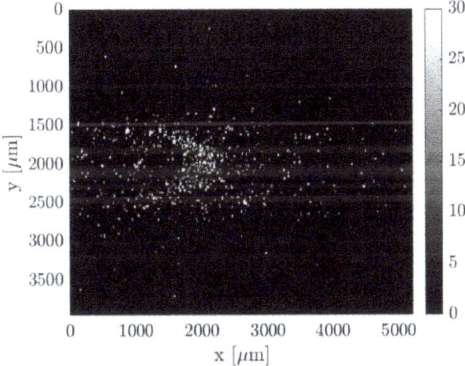

Figure 6. Example of a raw image acquisition of the acetone emission in argon flow at $p_m = 30$ kPa, $\Delta p = 550$ Pa, and an acetone molar concentration $\chi = 20\%$. The image results from the integration of $N_l = 100$ laser excitations, with delay $\Delta t = 50$ μs, gain $G = 100\%$, IRO gate $\Delta t_{gate} = 100$ ns, and 4×4 binning.

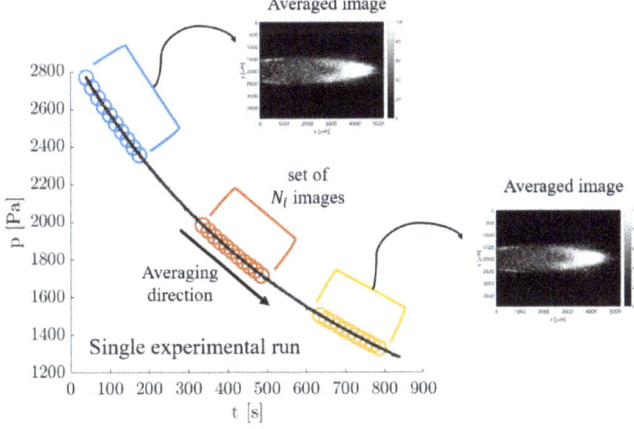

Figure 7. Qualitative illustration of the strategy used for producing an averaged image from a group of N_i images. The average is made "in cascade" on N_i images recorded during the same experimental run.

By recording images with the same delay Δt during one complete experimental run, the average could be applied to any group of N_i images that belongs to a time window centered around the experimental condition of interest. By moving the averaging window along the experimental run, a great number of averaged images that represent a multitude of thermodynamic conditions could be obtained from one single experiment. Thus, this averaging strategy could be considered to be not very time-consuming since a high number of averaged images at a specific delay Δt and for different average pressures and pressure differences could be obtained in a 30-min experimental run.

In order to obtain measurements at different delays, several experimental runs with the same initial boundary conditions need to be launched. This has to be done in order to capture the displacement profile of the same flow conditions but at different times after the laser excitation. For vacuum conditions at the outlet, the only parameter that needs to be regulated at the beginning of the experiment is the upstream pressure p_1. Therefore, the same initial boundary conditions could be easily reset for different experimental runs. The repeatability of the experimental conditions is important for both the mass flow rate measurements provided by the CV technique (see Section 5) and the velocity measurements provided by MTV. For the former technique, the possibility of reproducing the same experimental conditions allowed us to repeat multiple times the mass flow rate measurements, thus helping in characterizing its statistical uncertainty. For MTV, the fact that the same experimental flow can be reproduced with the highest accuracy is, instead, essential for comparing velocity measurements that come from displacement profiles recorded at different delays after the laser excitation. By repeating several times experimental runs with the same initial conditions of pressure and gas composition in the upstream tanks, it could be ascertained that the gas circuit used in this work was able to reproduce the same flow conditions along the whole duration of the experimental run and that no appreciable differences were present from one experiment to another.

The drawback of this averaging strategy is, however, the fact that the group of N_i images used for producing one averaged image corresponded to slightly different thermodynamic conditions. The group of images could not be, therefore, arbitrary large. The size of the averaging window needs to be small enough, so that the experimental conditions of the investigated flow vary only in an acceptable range. In this perspective, it is of fundamental importance to assess how much the experimental conditions vary from one image to the other. The MTV results presented in Section 6 were performed with N_i = 20, therefore the following analysis was presented only for this number of averaged images. By considering that the time between two consecutive image acquisitions was 5 s, the recording of N_i = 20 images took 150 s. In Figure 8, the velocity profiles together with their variations in respect to time were calculated at t = 100, 1000, and 2000 s from the beginning of a representative experimental run (the same as in Figure 5). The velocity profiles were computed from the slip flow analytical solution proposed by Ebert and Sparrow [19] by assuming a tangential momentum accommodation coefficient (TMAC) equal to 1.

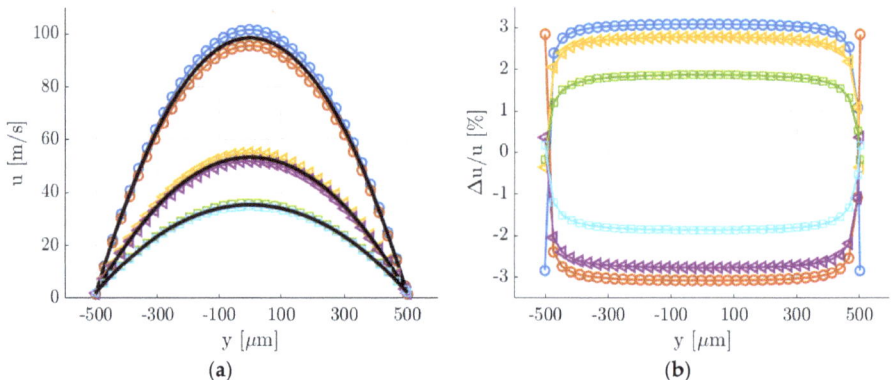

Figure 8. (a) Analytical velocity profiles $u(y,t)$ at t = 100 s (O), 1000 s (◁), and 2000 s (□) during the experimental run. The evolution of $u(y,t)$ during the recording of N_i = 20 consecutive images is illustrated. At each time t, the figure reports: the velocity profile captured by the first image of the group; the velocity profile captured by the last image of the group; the average velocity profile (solid black lines) that the image resulting from the average of the N_i images would represent. (b) Relative variation of the velocity profile with respect to the average velocity profile at t = 100 s (O), 1000 s (◁), and 2000 s (□).

At t = 0, that is for the highest pressure differences imposed and average pressures, the velocity variations with time were the highest, with a maximum value of about 3%. At t = 2000 s, that is for the lowest pressures, the velocity variations with time were the lowest, with a maximum variation of less than 2% at the velocity centerline. Consequently, the time stability of the velocity profile was lower at higher pressures. As the average pressure in the system decreased and the slip regime was approached, the velocity profile varied slower. Moreover, while the centerline velocity substantially decreased during one complete experimental run, the slip velocity at the wall was much more stable in time and varied from 1.6 m/s at t = 0 to 2.2 m/s at t = 2000 s. This peculiarity of the experimental flow conditions is of great help for the validation of MTV velocity reconstructions. As the Knudsen number at the test section increased, the ratio of the slip velocity to the average velocity was expected to raise from 2.5% to 10%.

4. MTV Image Post-processing

The main objective of the post processing procedure was to identify the position and the shape of the deformed tagged line. Several steps are necessary for accurately extracting the displacement profile from a noisy image. The flow chart in Figure 9 shows the sequence of post-processing operations, which are image averaging, wall detection, detection of the tagged line initial position, filtering of the background noise, Gaussian fitting per line, and extraction of the displacement data s_{xj}. In step (a), an example of image resulting from averaging N_i images is shown. Steps (b) and (c) are grouped, as they are operations that simply modify the size of the image or the Cartesian coordinate system origin. In step (d), MTV images are passed through a proper noise filter to make the fitting procedure of step (e) more robust. Finally, step (f) consists in the actual measurement of the velocity profile by means of the reconstruction method [13]. All the operations described below were implemented by means of Matlab® software (The MathWorks Inc., Natick, MA, USA) and were systematically applied to all the recorded images. The way step (a) is performed was thoroughly described in the previous section. The next steps of the procedure are discussed below in detail.

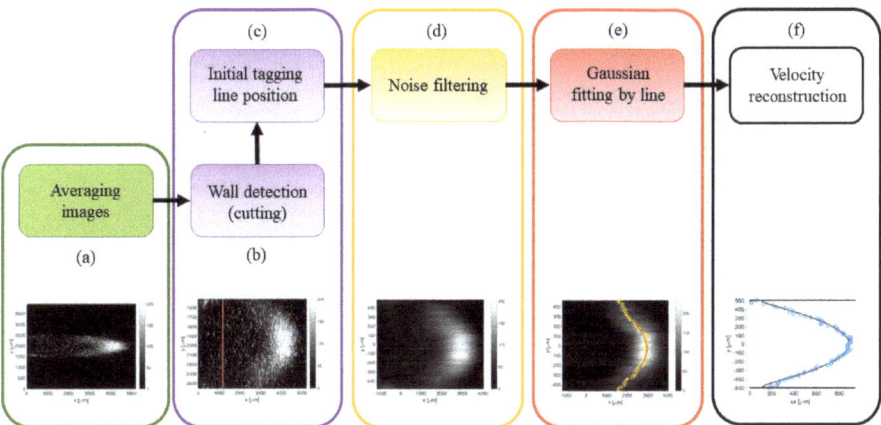

Figure 9. Flow chart of the image post-processing: (**a**) image averaging; (**b**) detecting wall position and image cutting; (**c**) detecting tagged line position; (**d**) filtering background noise; (**e**) Gaussian fitting per each horizontal line of pixels; and (**f**) velocity reconstruction from the displacement data s_{xj}.

(b) Wall detection: the strategy employed in this work for the identification of the channel walls was based on the investigation of the light emission provided by the top and bottom Suprasil® windows. The fluorescence of the silica is short-lived but very strong in the first few nanoseconds after the laser excitation. By recording the light emission of the silica with no tracer vapor inside the channel, the spots of maximum intensity on the windows correspond to the locations of entrance and exit of the laser beam, and, therefore, they represent the position on the image of the intersection between the tagged line and the channel walls. The procedure described above has a limited precision in localizing the channel walls. Firstly, the lowest possible uncertainty is determined by the pixel size. Secondly, the maximum intensity emission on the Suprasil® windows represents only an indication of where the wall is located. The identified wall lines may correspond to the solid surface that is just before or just after the plane where the tagged line is developing. Nevertheless, this strategy for detecting the wall position on the image is currently the most accurate one available.

(c) Initial tagged line position: the measurement of the displacement profile along the channel height requires first the identification of the initial position of the tagged molecules. In order to minimize possible displacements of the vertical tagged line from its original position, the flow rate needs to be low enough with respect to the chosen delay of acquisition. The interest in slightly delaying the time of acquisition with respect to the laser excitation derives from the idea of getting rid of the intense short-lived emission of Suprasil®. Based on the gas flow speed, the delay of acquisition needs to be chosen small enough to maintain the displacement of the tagged line at least smaller than 1 pixel. For a gas flow with an average velocity of 30 m/s, the delay cannot be higher than 100 ns or 500 ns, respectively whether no binning or a 4 × 4 binning is used. The post-processing procedure for the computation of the tagged line position is quite trivial. Figure 10 illustrates an example of early emission from the tagged molecules. Firstly, a region of interest is defined around the tagged line and, secondly, the position of the maximum light intensity is localized along the tagged line. A linear fitting of the light peaks provides the mean abscissa μ_x of the tagged line and a possible inclination of the laser beam with respect to the wall surface. Since the SNR is relatively high, it is preferred to avoid the use of binning for obtaining the highest precision in the evaluation of μ_x. By assuming that the laser beam position does not move with respect to the CCD, the abscissa μ_x is considered as the starting position of the tagged molecules for the images representing the deformed tagged line. The x-coordinate system of these images is translated so that the origin $x = 0$ corresponds to the initial position of the tagged line.

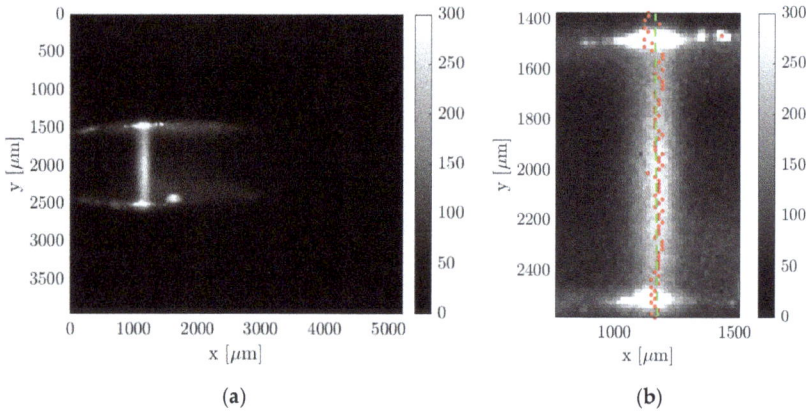

Figure 10. Post-processing procedure for detecting the initial position of the tagged line: (**a**) raw image of acetone early emission and (**b**) fitting of the maximum intensity along the y-coordinate detected in the region of interest.

(d) Noise filtering: the identification of the deformed tagged line is based on the application of appropriate fitting functions on the recorded image. The fitting procedure is a tool for extracting the most important information out of the background noise. Nevertheless, the SNR of the image is sometimes so low that the background noise prevents the convergence of the fitting optimization. In these cases, it is necessary to first pretreat the image with a noise filter for improving its SNR. A first strategy for reducing the background noise is based on the application of a 2D filter, which is a filter function applied on the whole image. The type of 2D filter employed in this work is the median filter, which is very efficient in removing the speckle noise. A second possibility consists of applying a 1D noise filter on each horizontal line of pixels. A family of digital filters that can be employed for this purpose is the finite impulse response (FIR) filters family. Both the 2D median filter and the 1D FIR filters have been used to improve the SNR of the image. However, the median filter has shown to be more efficient in improving the convergence of the fitting procedure while maintaining the sharpness of the tagged line.

(e) Gaussian fitting per line: in order to track the molecular displacement at different times from the initial position, an appropriate criterion for identifying the tagged line profile is required. Previous analyses have shown that the best fitting function that represents the evolution in time of the light distribution is the Gaussian function [15]. Nevertheless, in a confined gas flow, the light distribution may be affected by the combined mechanisms of streamwise advection and spanwise molecular diffusion in the direction perpendicular to the wall, thus differing from the Gaussian distribution in a quiescent gas. However, the image data revealed that these effects are negligible, and the Gaussian function is still representative of the light distribution in the gas flow direction. As it has been done in previous works [12,14,20], the Gaussian fitting applied to each horizontal line of pixels at any position $y = y_j$ is

$$f(x) = \frac{a_2}{\sqrt{2\pi a_3}} e^{-\frac{(x-a_4)^2}{2a_3}} + a_1, \qquad (1)$$

where the fitting parameters a_1, a_2, a_3, and a_4 represent, respectively, the offset, the amplitude, the variance, and the peak position. Thus, $a_{4,j}$ corresponds to the displacement $s_{x,j}$ at each discrete location y_j. Figure 11 illustrates the procedure of the Gaussian fitting function applied to a horizontal line of pixels on an MTV image. As it can be observed, the statistical fluctuations characterizing the signal distribution on one horizontal line of pixels were significant with respect to the Gaussian peak. This certainly increased the uncertainty in the localization of the tagged line profile. Since the whole displacement data $s_{x,j}$ along the channel height will be afterwards fitted by the numerical solution $s_x(y_j)$ of the advection–diffusion equation, the uncertainty on the tagged line position at each location y_j could be practically quantified by the ratio of the deviation between the displacement data and the numerical displacement profile, $s_{x,j} - s_x(y_j)$, to the value of the displacement profile $s_x(y_j)$ itself.

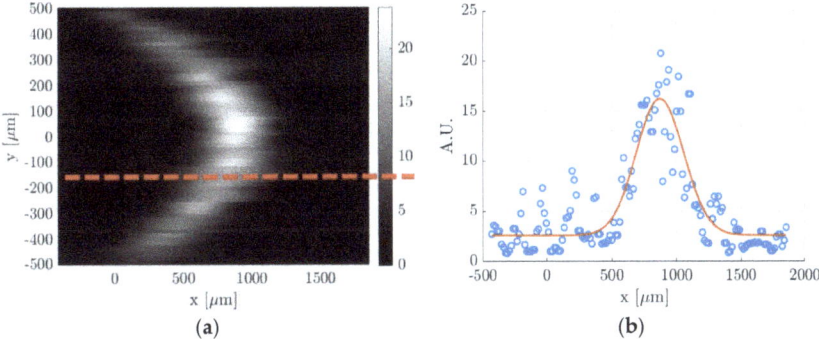

Figure 11. (a) Image resulting from the application of the Gaussian fitting per line to a raw MTV image and (b) example of data distribution along a horizontal line of pixels, indicated by the red dashed line in (a), and the corresponding Gaussian fitting (A.U. stands for arbitrary unit).

(f) Velocity reconstruction: the final step is the application of the velocity reconstruction algorithm to the set of displacement data $s_{x,j}$ determined in step (e). The reconstruction method of Frezzotti et al. [13] provides the velocity profile that makes the numerical displacement $s_x(y)$ fit at best the displacement dataset $s_{x,j}$. The numerical solving of the diffusion–advection equation requires providing the channel height H and the diffusion coefficient D characterizing the tracer diffusion through the background gas. The reconstruction method is sensitive to both parameters. For a given velocity profile $u(y)$ and a given diffusion coefficient D, the displacement profile $s_x(y)$ at a certain time t has a thickness, defined as $\Delta s = \max s_x(y) - \min s_x(y)$, that is proportional to H^2 [13]. The value of the wall distance H used in defining the mathematical domain needs, therefore, to be as accurate as possible in order to have a correct velocity reconstruction. As discussed in step (b), the localization of the channel walls on the image cannot be done with the highest accuracy and is affected by possible imperfections in the calibration of the positioning of the channel and the laser beam with respect to the CCD. For flow regimes characterized by strong molecular diffusion, the numerical solution $s_x(y)$ is also sensitive to the value of the diffusion coefficient D.

5. CV Methodology and Data Processing

The local measurements from the molecular tagging are compared to global measurements obtained by means of the constant volume technique. The CV technique allows one to relate pressure variations with time to mass variation with time inside a reservoir of constant volume. Thus, monitoring the time evolution of the pressure in the upstream tank allows one to measure the mass flow rate $\dot{m}(t)$ at any time during the experimental run [21]. If the pressure variation due to thermal fluctuations is neglected, the equation relating pressure variation to mass variation is

$$\dot{m}_{CV} = -\frac{V}{R_s T}\frac{dp_1}{dt}. \qquad (2)$$

where R_s is the specific gas constant, which the value depends on the acetone concentration present in the gas mixture. The gas temperature T was assumed to correspond to the regulated temperature (24 °C) in the experimental room. The upstream reservoir corresponds to the region in Figure 2 encompassed by the dashed line, and it is composed of the two tanks and all valves and connection pipes that are upstream valve V1. The methodology used for measuring the volume V of the upstream reservoir consists of a simple thermodynamic experiment: (i) with valves V1 and V5 closed, the upstream reservoir is first vacuumed through valve V12 down to a pressure p_v; (ii) valve V12 is then closed and a small reference reservoir filled with air at atmospheric pressure p_{atm} is connected to the vacuumed reservoir in correspondence of valve V12; and (iii) by opening valve V12, the air in the small reservoir is discharged into the upstream reservoir until an equilibrium pressure p_{eq} is reached. The volume of the small reservoir is known to be $\overline{V} = 486.5 \times 10^{-6}$ m^3 and it has been accurately calculated by measuring its weight when filled with water. By considering that at the initial and final thermodynamic states the air is at ambient temperature, the volume of the reservoir is calculated from

$$p_v V + p_{atm}\overline{V} = p_{eq}(V + \overline{V}). \qquad (3)$$

This experiment has been repeated several times for a statistical characterization of the measurement uncertainty. The volume of the upstream reservoir has been measured to be $V = 0.176$ m^3 with a relative uncertainty of 0.08%. The evaluation of the derivative $\frac{dp_1}{dt}$ is carried out by fitting the pressure data $p_1(t)$ with a proper analytical function. The most suitable fitting function for the pressure relaxation characterizing the type of experimental setup employed in this work is a series of N exponential functions:

$$p_1 = \sum_{i=1}^{N} a_i e^{-t/\tau_i}. \qquad (4)$$

The choice of N is based on the length of the time interval of pressure data to be fit. For long time intervals, more exponential functions are used for obtaining the best fittings. For the experimental run of Figure 5, $N = 4$ provided an accurate fitting of the whole array of pressure data. Once the fitting parameters a_i and τ_i have been determined, the rate of variation of the upstream pressure and the mass flow rate entering in the channel can be calculated.

From the measured inlet and outlet pressures p_1 and p_2, the axial pressure distribution $p(x,t)$ along the channel can be predicted from the analytical solution obtained by Arkilic et al. [22]. Figure 12 reports $p(x,t)$ at three different times $t = 100, 1000,$ and 2000 s from the beginning of the experiment.

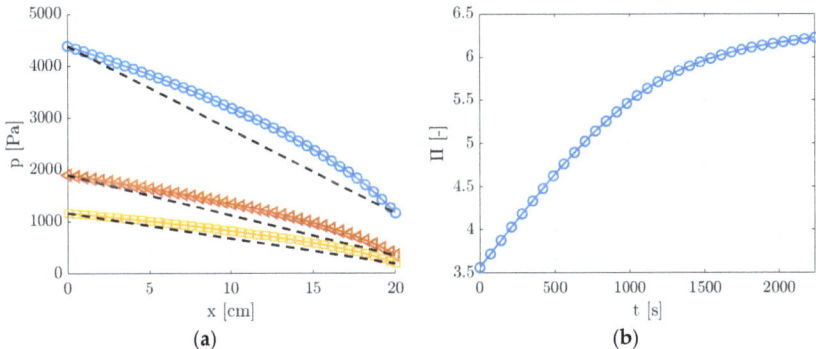

Figure 12. (a) Pressure distribution $p(x,t)$ along the channel length at three different times t = 100, 1000, and 2000 s during the experimental run, which correspond to pressure ratios Π = 3.8 (O), 5.5 (◁), and 6.2 (□). The dashed lines represent the linear pressure distributions when rarefaction and compressibility effects are neglected; (b) time evolution of pressure ratio Π during the experimental run.

Since the pressure ratio $\Pi = p_1/p_2$ was relatively high, ranging from 3.5 to 6.5, the compressibility effects on the pressure distribution were not negligible for the whole duration of the experimental run. As shown in Figure 12b, even though the pressure difference $p_1 - p_2$ decreased in time, the pressure ratio Π increased and, thus, the effects of the gas compressibility on the pressure distribution along the channel assumed even more importance at the lowest pressures.

The mass flow rate $\dot{m}_{CV}(t)$ provided by the CV technique and the theoretical pressure distribution $p(x,t)$ inferred from the pressure measurements $p_1(t)$ and $p_2(t)$ provided an estimation of the average velocity of the flow on the cross-section, $\bar{u}_{CV,2D}(x,t)$, at each position x along the channel and at each time t during the experimental run:

$$\bar{u}_{CV,2D}(x,t) = \frac{\dot{m}_{CV}(t)\, R_s T}{Hb\, p(x,t)}. \quad (5)$$

Nevertheless, $\bar{u}_{CV,2D}$ represents the average of the velocity over the entire cross-section of the channel, while the average velocity $\bar{u}_{MTV}$ measured by MTV represents the average value along a line at $z = 0$. Therefore, in order to have a meaningful comparison between MTV and CV data, the average value $\bar{u}_{CV,1D}$ of the velocity along this line was computed from the average value $\bar{u}_{CV,2D}$ on the section by using the analytical solution of the 2D velocity distribution over the cross-section provided by Ebert and Sparrow [19]. The analytical 2D velocity function was evaluated at $z = 0$ and averaged over the channel height, yielding the following relationship between $\bar{u}_{CV,1D}$ and $\bar{u}_{CV,2D}$:

$$\bar{u}_{CV,1D}(x,t) = K(x,t)\bar{u}_{CV,2D}(x,t). \quad (6)$$

By setting the TMAC to 1, the value of the coefficient $K(x,t)$ depends on the Knudsen number of the gas mixture and was therefore computed for each thermodynamic condition. Finally, the CV measurement that was compared to the average velocity $\bar{u}_{MTV}$ measured by MTV corresponded to the value of $\bar{u}_{CV,1D}(x,t)$ evaluated at $x = L/2$, which, for the sake of simplicity, will be just indicated as $\bar{u}_{CV}$. MTV measurements of the velocity profile might not fall exactly at $z = 0$ along the channel width but the variations of the profile are less than 1% even 1 mm apart from the center. The comparison between MTV and CV velocity measurements would provide a validation and verification of the two experimental techniques and of the analytical solutions for the pressure distribution and for the 2D velocity profile of Ebert and Sparrow, which were both used for the calculation of $\bar{u}_{CV}$.

Although the gas compressibility produces non-negligible variations of the flow properties along the channel length (Figure 12a), the investigated gas flow could always be assumed as locally fully developed, locally incompressible, and isothermal. The analytical solution for the pressure distribution provided by Arkilic et al. [22] can be employed only if these assumptions can be considered as valid. Harley et al. [23] compared the analytical solution based on these same hypotheses with the numerical solution of the full compressible Navier–Stokes equations for a slip gas flow in a microchannel. They reported discrepancies of less than 3% between the theoretical and numerical solutions of the axial velocity profile for local Mach numbers lower than 0.3. For the same range of Mach number values, the numerical results in [23] showed that the temperature drop due to the gas expansion along the channel is less than 3%. From the results of Equation (5) and by assuming an isothermal gas flow, the Mach number $Ma(x,t)$ of the gas flow here investigated was calculated to be lower than 0.3 at any position along the channel length and at any time during an experimental run.

6. Experimental Results

The presentation of the experimental data was organized in the following manner. Section 6.1 reported MTV measurements of argon flows in non-rarefied conditions and at pressures between 30 and 6 kPa. Section 6.2 presented MTV data of helium flow in rarefied conditions and at pressures between 1.6 kPa and 900 Pa. In Section 6.3, the comparison between MTV and CV data was analyzed for the cases presented in Sections 6.1 and 6.2. The kinetic diameter for the acetone molecule used for all the calculations was $d = 470$ pm, as reported in Van der Perre et al. [24], which is also similar to the values reported in other recent works [25,26]. Moreover, since the investigated gas flow can be considered as isothermal, the mean free path of the mixture was estimated by modeling molecular interactions through hard sphere potentials.

6.1. MTV Measurements of Channel Gas Flows in Non-rarefied Conditions

In this section, the MTV technique was applied to acetone-seeded argon flows in non-rarefied conditions. The previous works of Samouda et al. [12] and Si-Hadj Mohand et al. [14] already demonstrated the successful application of MTV for velocity measurements in channel gas flow in non-rarefied conditions by using as a tracer acetone vapor excited at 266 nm. Si Hadj Mohand et al. were not able to apply MTV for pressures lower than about 42 kPa because of the low signal intensity due to the limited phosphorescence quantum yield of acetone excited at 266 nm. The results of this work were based on acetone vapor excited at 310 nm, which guaranteed a detectable and longer signal even for pressures lower than 1 kPa.

In order to demonstrate that the novel experimental setup provided accurate results in non-rarefied conditions, the first results presented in this work are for gas flows at pressures between 30 and 6 kPa. These are the first MTV results obtained for gas flows at pressures lower than 42 kPa. In these experiments, the mass flow rate, and thus the inlet over outlet pressure ratio were maintained low in order to make the upstream pressure vary slowly enough and to obtain quasistationary conditions during one experimental run. For the thermodynamic conditions considered in this section, the Mach number was at most 0.04, and the Knudsen number was not higher than 0.001. For this reason, the local compressibility and the rarefaction effects were negligible. In addition, the pressure variation along the channel was low (inlet–outlet pressure difference between 2% and 5% of the average pressure), and, consequently, the theoretical pressure distribution along the channel was quasilinear.

Figures 13 and 14 show the molecular displacement of the tagged line through the channel, at different delays after laser excitation, from $\Delta t = 10$ to 50 µs. Each image represents the average of 20 images, each integrating 100 laser pulses. The acquisitions were carried out by setting the gain at its maximum value, i.e., $G = 100\%$, the IRO gate at minimum, i.e., $\Delta t_{gate} = 100$ ns, and by using a 4 × 4 binning. A 2D median filter was applied to each image for increasing its SNR. The average pressure and the inlet–outlet pressure difference were quite similar for images acquired at different delays Δt. Therefore, we considered that the comparison of the molecular displacement at different Δt was still representative of how the tagged line evolves as it moves through the channel. Average pressure p_m, inlet–outlet pressure difference Δp, mass flow rate $\dot{m}_{CV}$, Mach number Ma, and Knudsen number Kn that characterize each image are summarized in Tables 1 and 2. As previously mentioned, since the experiments were conducted in quasistationary conditions, the thermodynamic conditions in the system varied with time. The tables reported the variations in percentage with respect to the average value of each property during 100 s, which is the time needed to acquire $N_i = 20$ images.

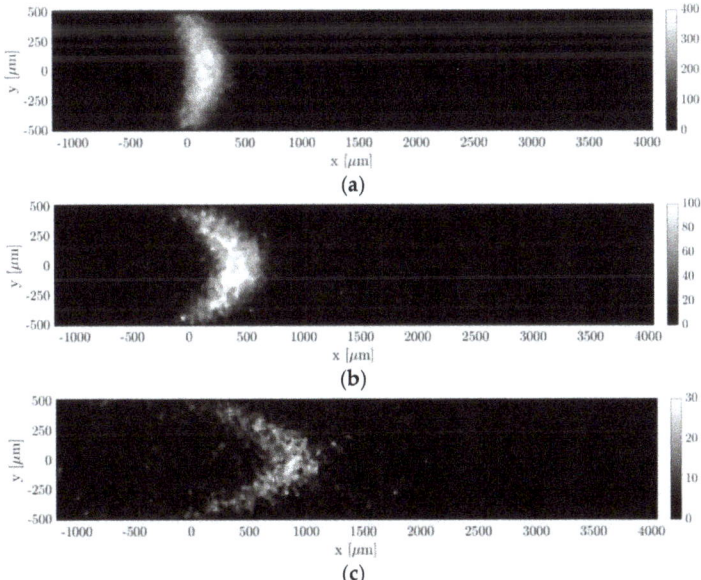

Figure 13. MTV acquisitions of argon–acetone flow with $\chi = 20\%$: (**a**) $\Delta t = 10$ μs, $p_m = 27.38$ kPa, and $\Delta p = 528.4$ Pa; (**b**) $\Delta t = 25$ μs, $p_m = 28.83$ kPa, and $\Delta p = 538.8$ Pa; and (**c**) $\Delta t = 50$ μs, $p_m = 30.36$ kPa, and $\Delta p = 549.9$ Pa. The recording parameters are $N_l = 100$, $N_i = 20$, $G = 100\%$, and $\Delta t_{gate} = 100$ ns. The acquisitions are made with a 4 × 4 binning. The laser pulse energy was varying between 30 and 40 μJ.

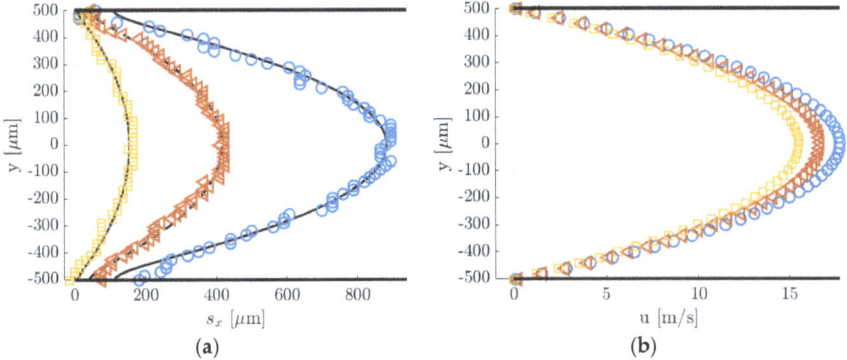

Figure 14. (**a**) Displacement data computed by applying the Gaussian fitting per line to the images of Figure 13. The black lines represent the numerical solutions for the displacement profile provided by the reconstruction method; (**b**) velocity reconstruction from the displacement data in (a): $\Delta t = 10$ μs (□), 25 μs (◁), and 50 μs (O).

Table 1. Flow properties associated to the images of Figure 13. The reported percentage change indicates how much the properties vary inside the group of $N_i = 20$ averaged images.

Image	Δt (µs)	p_m (kPa)	Δp (Pa)	$\dot{m}_{CV}$ (kg/s)	Ma (–)	Kn (–)
Figure 13a	10	27.38 ± 1.5%	528.4 ± 0.6%	2.63 × 10^{-5} ± 1.6%	0.039 ± 0.15%	1.85 × 10^{-4} ± 1.5%
Figure 13b	25	28.83 ± 1.5%	538.8 ± 0.6%	2.79 × 10^{-5} ± 1.6%	0.039 ± 0.15%	1.75 × 10^{-4} ± 1.5%
Figure 13c	50	30.36 ± 1.5%	549.9 ± 0.6%	2.95 × 10^{-5} ± 1.6%	0.039 ± 0.15%	1.66 × 10^{-4} ± 1.5%

Table 2. Flow properties associated to the images of Figure 15. The reported percentage change indicates how much the properties vary inside the group of $N_i = 20$ average images.

Image	Δt (µs)	p_m (Pa)	Δp (Pa)	$\dot{m}_{CV}$ (kg/s)	Ma (–)	Kn (–)
Figure 15a	20	6904 ± 1.2%	371 ± 0.3%	5.4 × 10^{-6} ± 1.5%	0.032 ± 0.27%	7.3 × 10^{-4} ± 1.2%
Figure 15b	30	6620 ± 1.2%	367 ± 0.31%	5.2 × 10^{-6} ± 1.5%	0.031 ± 0.29%	7.3 × 10^{-4} ± 1.2%
Figure 15c	40	6337 ± 1.2%	362.8 ± 0.31%	4.9 × 10^{-6} ± 1.5%	0.031 ± 0.3%	8 × 10^{-4} ± 1.2%
Figure 15d	50	6071 ± 1.2%	358.6 ± 0.31%	4.6 × 10^{-6} ± 1.5%	0.031 ± 0.31%	8.3 × 10^{-4} ± 1.2%

6.1.1. Experiments at High Pressure (p_m = 29 kPa)

Hereafter we presented the results of an argon–acetone gas mixture with χ = 20% acetone molar concentration. The average mean pressure of the gas flow was p_m = 28.85 kPa and the average Knudsen number was Kn = 1.75 × 10^{-4}. Figure 13 shows how the molecular displacement of acetone inside the background gas evolved over time. Two physical phenomena are immediately recognizable in the images, which are the advection and diffusion of the tracer inside the gas flow. As the delay from the laser excitation increased, it is evident from the images how the signal intensity reduced.

In Figure 14a, the data points represent the displacement data $s_{x,j}$ obtained by applying to the raw images of Figure 13 the Gaussian fitting per line procedure described in Section 4. The application of the reconstruction method by Frezzotti et al. [13] to these data provides the velocity profiles shown in Figure 14b. The black lines reported in Figure 14a represent the numerical solutions provided by the advection–diffusion equation of the molecular displacement generated by the velocity profiles of Figure 14b. The reconstruction algorithm was successful in accurately extracting the distribution of the axial velocity. As expected, the slip velocity at the wall was very close to zero, with values ranging between 0.006 and 0.2 m/s. The diffusion coefficient D employed for the velocity reconstruction has been estimated by means of the Blanc's law [15] and was set to 3.5 × 10^{-5}, 3.3 × 10^{-5} and 3.2 × 10^{-5} m^2/s, respectively, for the data at Δt = 10, 25, and 50 µs. Since argon and acetone have molecular mass and kinetic diameter of the same order of magnitude, small variations of the tracer concentration χ in different experimental runs did not modify considerably the estimated diffusion coefficient used in the reconstruction. Moreover, preliminary tests showed that for these experimental conditions the resulting velocity profile was quite insensitive to the value of the diffusion coefficient used.

6.1.2. Experiments at Low Pressure (p_m = 6.5 kPa)

Argon–acetone mixture flows were also investigated by MTV at lower pressures, in order to verify the reconstruction method in non-rarefied conditions but with higher molecular diffusion. The images of Figure 15 illustrate MTV acquisitions carried out at an average pressure of 6.5 kPa and at Δt = 20, 30, 40, and 50 µs. All thermodynamic characteristics of the observed experimental flow are reported in Table 2. As for the case at higher pressure, the thermodynamic properties associated to each image were not exactly the same but, again, quite similar. Since the gas pressure was 4 times lower than the case presented previously, the Knudsen number was 4 times higher.

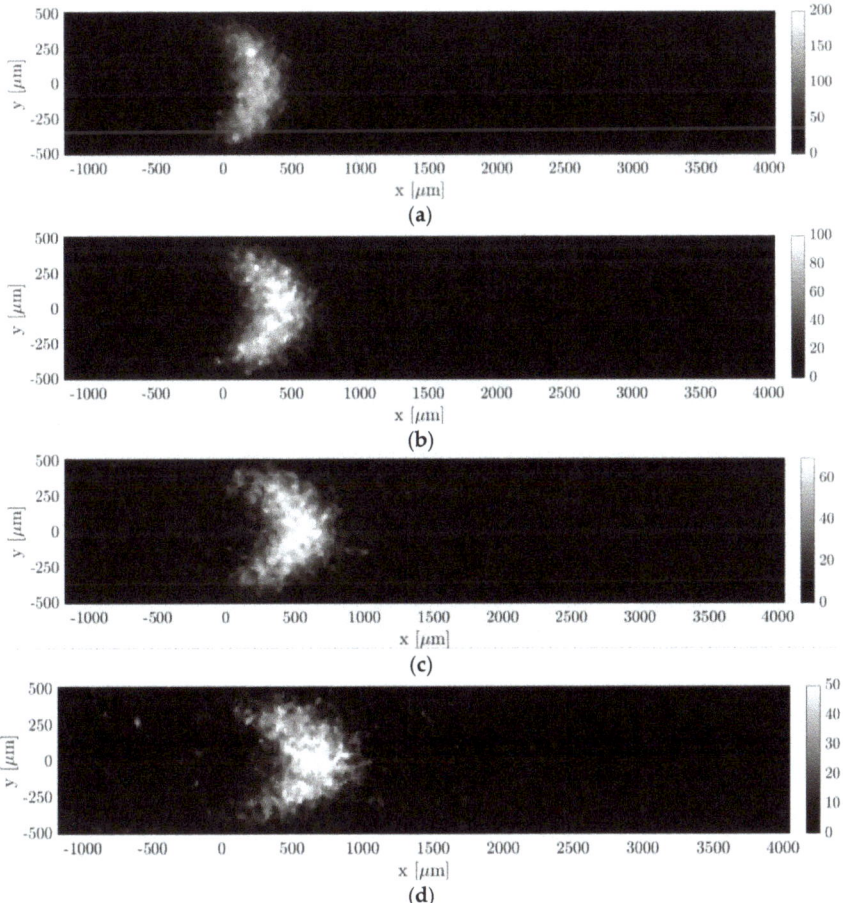

Figure 15. MTV acquisitions of argon–acetone flow with $\chi = 20\%$: (**a**) $\Delta t = 20$ μs, $p_m = 6904$ Pa, and $\Delta p = 371$ Pa; (**b**) $\Delta t = 30$ μs, $p_m = 6620$ Pa, and $\Delta p = 367$ Pa; (**c**) $\Delta t = 40$ μs, $p_m = 6337$ Pa, and $\Delta p = 362.8$ Pa; and (**d**) $\Delta t = 50$ μs, $p_m = 6071$ Pa, and $\Delta p = 358.6$ Pa. The recording parameters are: $N_l = 100$, $N_i = 20$, $G = 100\%$, and $\Delta t_{gate} = 100$ ns. The acquisitions are made with a 4×4 binning. The laser pulse energy was varying between 20 and 30 μJ.

Figure 16a,b reports, respectively, the displacement data with the corresponding numerical displacement profiles and the reconstructed velocity profiles. The diffusion coefficients used for the reconstructions were 1.39×10^{-4}, 1.46×10^{-4}, 1.52×10^{-4}, and 1.59×10^{-4} m^2/s, respectively, for the data at $\Delta t = 20$, 30, 40, and 50 μs. As it can be observed, the reconstruction method was not able to provide here an accurate velocity profile. The slip velocity at the wall, which was expected to be limited to a maximum of 0.05 m/s, ranged between 1 and 2.6 m/s, depending on the delay of the displacement data used. Figure 16a shows that the displacement data were well fitted by the numerical solution provided by the reconstruction method. Except for the case at $\Delta t = 50$ μs, the dispersion of displacement data around the numerical solution was quite low. However, while the relatively high SNR at the channel centerline $y = 0$ made the identification of the deformed tagged line quite accurate, there might be an uncertainty on the local displacement close to the wall. With respect to the displacement profiles at 28 kPa previously reported, the molecular displacement at the wall was, in this case, definitely larger as a consequence of the higher molecular diffusion. The accurate evaluation of the molecular slip at the wall appeared to be of fundamental importance for extracting a correct velocity profile.

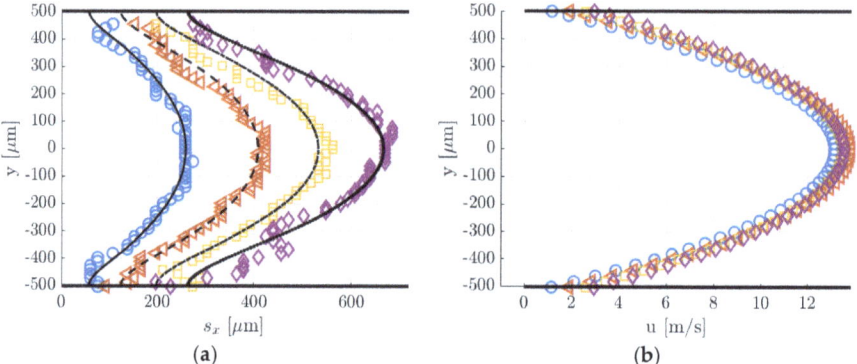

Figure 16. (a) Displacement data computed by applying the Gaussian fitting per line to the images of Figure 15. The black lines represent the numerical solutions for the displacement profile provided by the reconstruction method; (b) velocity reconstruction from the displacement data in (a): Δt = 20 μs (○), 30 μs (◁), 40 μs (□), and 50 μs (◇).

6.2. MTV Measurements of Channel Gas Flows in Rarefied Conditions

In this section, MTV was applied to helium–acetone mixture flow with χ = 20% in the slip regime. Two experimental conditions of this gas flow were here investigated. The first one was at an average pressure of about 1.6 kPa, the second one was at 920 Pa. All properties of the visualized gas flow are reported in Tables 3 and 4, along with their range of values that characterized the images used for averaging. Figures 17 and 18 illustrate the tracer displacement, respectively for the gas flow at 1.6 kPa and at 920 Pa, for four different delays after the laser excitation, at Δt = 20, 30, 40, and 50 μs. These images are the first flow visualizations ever obtained in confined and rarefied gas flows. They represent the experimental demonstration of the molecular displacement in a channel theoretically predicted by the diffusion–advection equation [13]. Due to the high repeatability of the experimental flow that the gas circuit can guarantee at low pressures, the four images corresponded to the very same average thermodynamic conditions. The Mach and the Knudsen numbers were on average, respectively, Ma = 0.07 and Kn = 0.008 for the case at 1.6 kPa, and Ma = 0.045 and Kn = 0.014 for the case at 920 Pa. The two investigated gas flows could then be considered as locally incompressible and at the beginning of the slip flow regime. Even though the Knudsen number of the gas flow at 920 Pa was almost twice as much as at 1.6 kPa, the slip velocity at the wall was estimated to be almost the same, 2.2 m/s and 2 m/s, respectively. This is because the flow speed imposed by the vacuum pumps decreased as the working pressure decreased.

The pressure difference and mass flow rate were, respectively, 1790 Pa and 1.7×10^{-6} kg/s for the images of Figure 17, and 1092 Pa and 1.7×10^{-7} kg/s for the images of Figure 18. The flow rate of these flows was higher than the one characterizing the argon–acetone flows, as it can be noted by comparing the imaged molecular displacements of Figures 17 and 18 with those of Figures 14 and 16. As for the previous case of argon–acetone flow, each image of Figures 17 and 18 represents the average of N_i = 20 images. However, because of the higher flow rates, the relative variations of the thermodynamic properties associated to a group of N_i averaged images were higher than for the data of argon flows. Nevertheless, even though the mass flow rate varied up to 6.5%, the variations on the velocity profile were limited to less than 3%, as previously discussed in Section 3.

Table 3. Flow properties associated to the images of Figure 17. The reported percentage change indicates how much the properties vary inside the group of $N_i = 20$ average images.

Image	Δt (µs)	p_m (Pa)	Δp (Pa)	$\dot{m}_{CV}$ (kg/s)	Ma (–)	Kn (–)
Figure 17a	20	1576 ± 3.6%	1789.2 ± 3.2%	1.7×10^{-6} ± 6.5%	0.07 ± 2.7%	0.008 ± 3.5%
Figure 17b	30	1580 ± 3.6%	1793.2 ± 3.2%	1.7×10^{-6} ± 6.2%	0.07 ± 2.5%	0.008 ± 3.5%
Figure 17c	40	1576.6 ± 3.6%	1789.2 ± 3.2%	1.7×10^{-6} ± 6.4%	0.07 ± 2.7%	0.008 ± 3.5%
Figure 17d	50	1578.7 ± 3.6%	1791.8 ± 3.2%	1.7×10^{-6} ± 6%	0.07 ± 2.2%	0.008 ± 3.5%

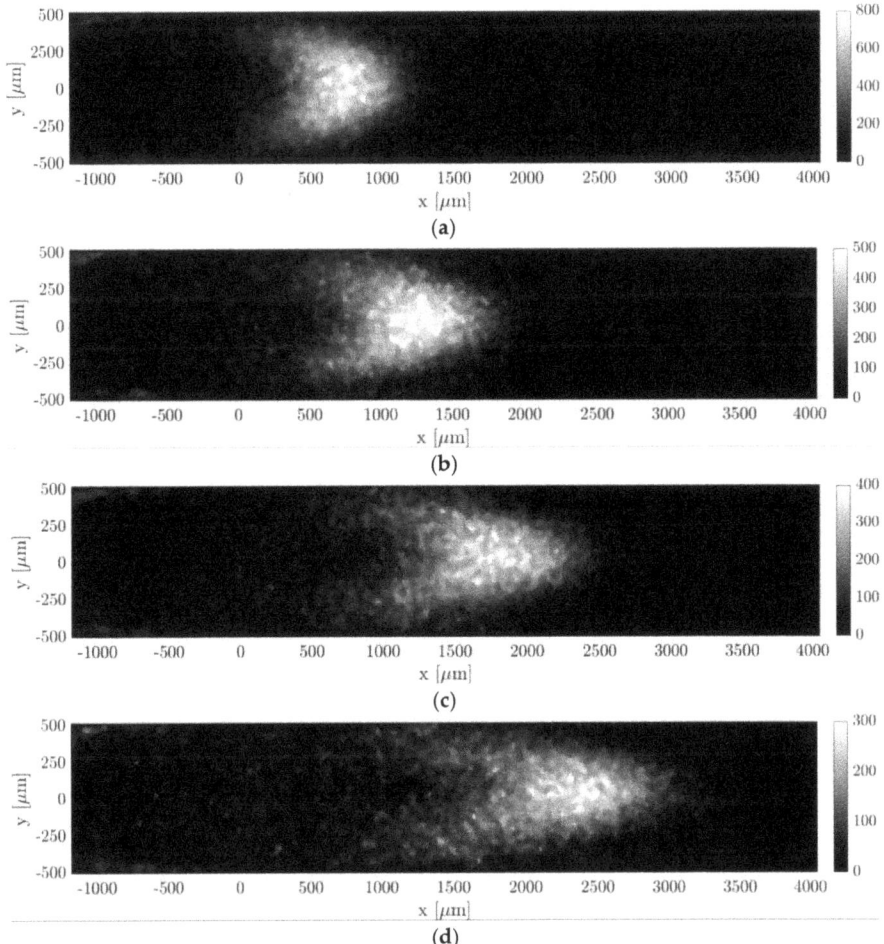

Figure 17. MTV acquisitions of helium–acetone flow with $\chi = 20\%$: (**a**) $\Delta t = 20$ µs, $p_m = 1576$ Pa, and $\Delta p = 1789.2$ Pa; (**b**) $\Delta t = 30$ µs, $p_m = 1580$ Pa, and $\Delta p = 1793.2$ Pa; (**c**) $\Delta t = 40$ µs, $p_m = 1576.6$ Pa, and $\Delta p = 1789.2$ Pa; and (**d**) $\Delta t = 50$ µs, $p_m = 1578.7$ Pa, and $\Delta p = 1791.8$ Pa. The recording parameters used are: $N_l = 100$, $N_i = 20$, $G = 100\%$, and $\Delta t_{gate} = 500$ ns. The acquisitions are made with a 4 × 4 binning. The laser pulse energy was varying between 80 and 120 µJ.

Table 4. Flow properties associated to the images of Figure 18. The reported percentage change indicates how much the properties vary inside the group of N_i = 20 average images.

Image	Δt (µs)	p_m (Pa)	Δp (Pa)	$\dot{m}_{CV}$ (kg/s)	Ma (–)	Kn (–)
Figure 18a	20	920 ± 2.3%	1092.7 ± 2.2%	6.5×10^{-7} ± 4.4%	0.045 ± 2%	0.014 ± 2.3%
Figure 18b	30	920.3 ± 2.3%	1093.4 ± 2.3%	6.5×10^{-7} ± 4%	0.045 ± 1.7%	0.014 ± 2.3%
Figure 18c	40	919.1 ± 2.3%	1091.8 ± 2.2%	6.5×10^{-7} ± 3.9%	0.045 ± 1.6%	0.014 ± 2.3%
Figure 18d	50	919.1 ± 2.4%	1090.7 ± 2.2%	6.5×10^{-7} ± 4.4%	0.045 ± 2%	0.014 ± 2.3%

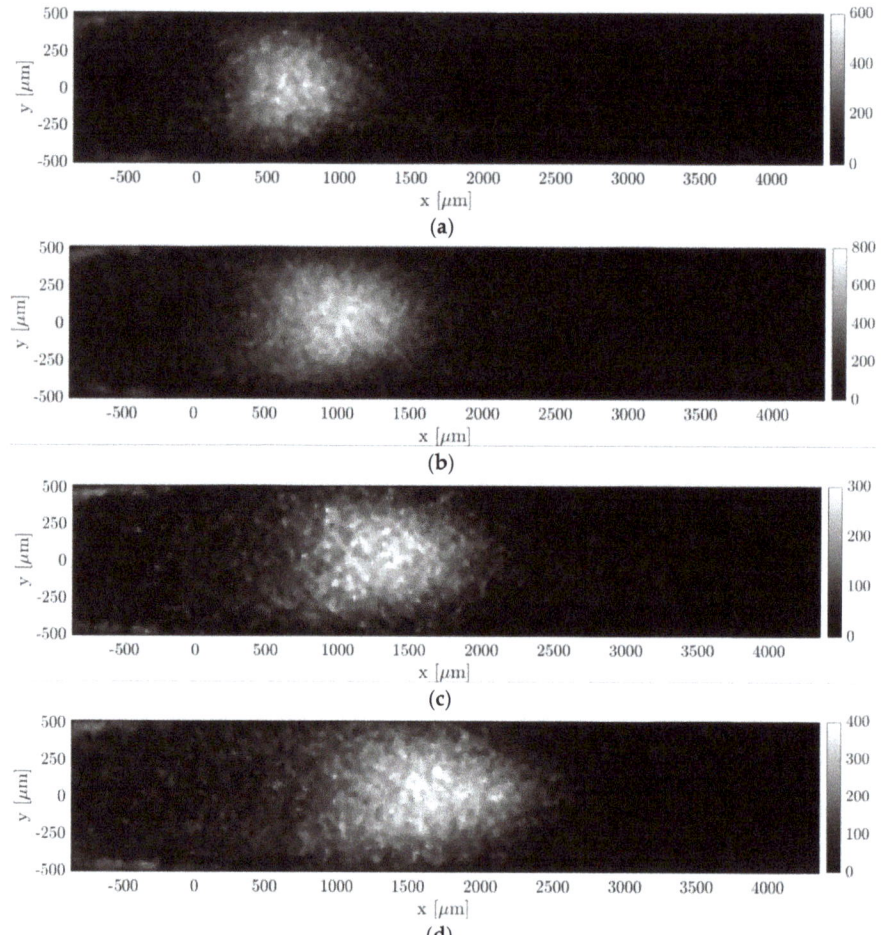

Figure 18. MTV acquisitions of helium–acetone flow with χ = 20%: (**a**) Δt = 20 µs, Δt_{gate} = 500 ns, p_m = 920 Pa, and Δp = 1092.7 Pa; (**b**) Δt = 30 µs, Δt_{gate} = 1000 ns, p_m = 920.3 Pa, and Δp = 1093.4 Pa; (**c**) Δt = 40 µs, Δt_{gate} = 500 ns, p_m = 919.1 Pa, and Δp = 1091.8 Pa; and (**d**) Δt = 50 µs, Δt_{gate} = 1000 ns, p_m = 919.1 Pa, and Δp = 1090.7 Pa. The recording parameters used are: N_l = 100, N_i = 20, and G = 100%. The acquisitions are made with a 4 × 4 binning. The laser pulse energy was varying between 80 and 100 µJ.

In all images of Figures 17 and 18, a peculiar noise was observed behind the wake of the moving emitting molecules, where the origin was still unclear. It is possible that this emission was provided by tagged molecules that have diffused from the center towards the lateral walls of the channel, where the gas velocity slows down.

By comparing the images of Figures 17 and 18 with those of Figures 13 and 15, it can be observed that the light emission was stronger for the data related to acetone–helium flow, even though the gas pressure was lower. This is because the IRO gate Δt_{gate} was increased and the average energy of the laser excitation was raised to values between 80 and 120 µJ. For the case at 1.6 kPa, Δt_{gate} was set to 500 ns, while for the gas flow at 920 Pa, IRO gates of 500 ns and 1000 ns were both used to investigate whether this acquisition parameter could improve or not the SNR of the MTV images. In particular, Δt_{gate} = 500 ns was used for the images at Δt = 20 and 40 µs, whereas Δt_{gate} = 1000 ns was used for the images at Δt = 30 and 50 µs. Doubling the IRO gate doubled the amount of light collected on the CCD, as expected, but it did not seem to drastically improve the SNR of the acquisition.

The images of Figure 18 anticipate which are the difficulties that will be encountered in applying MTV to gas flows at pressures even lower than those here investigated. As the working pressure decreased, the highest mass flow rate imposed by the two downstream vacuum pumps decreased as well. The reduced gas flow speed combined with the increased molecular diffusion makes the Taylor dispersion become very strong, which significantly flattens the profile of the tagged line. The increasing strength of the Taylor dispersion can already be noticed by comparing the MTV images of Figure 17 with those of Figure 18. At 1.6 kPa, the tagged line was deformed in time by the gas movement, and despite the relatively high molecular diffusion the CCD sensor could still clearly display the shape of the displacement profile. For the gas flow at 920 Pa, the Taylor dispersion and the molecular diffusion were starting to dominate over the gas advection and the tagged line was tending to become a cloud of molecules that migrated in the flow direction without preserving a defined shape. The current MTV setup was able to resolve the displacement profiles even at pressures as low as 920 Pa, but it might and probably failed in providing local information on the gas molecular displacement in flows at lower pressures. However, this limitation of the current experimental setup could be overcome by (i) increasing the CCD resolution, e.g., by reducing the binning or by increasing the magnification of the lenses system, and/or by (ii) adding more vacuum pumps at the channel outlet. The former solution would allow us to resolve smaller deformations of the tagged line, but it would have the drawback of reducing the amount of light collected and the SNR of the image data. The latter option would, instead, increase the speed of the gas flow thus decreasing the effects of the Taylor dispersion on the displacement profile. Increasing the flow speed would also favorably increase the magnitude of the slip velocity at the wall characterizing the observed gas flow. Nevertheless, in order to maintain the gas flow in a local-incompressibility regime, the Mach number needs to be kept low enough, and so the downstream pumping force cannot be raised too much.

Figure 19 reports the displacement data along with the numerical displacement profiles that result from the application of the reconstruction method. Since the displacement data at different delays corresponded to the same thermodynamic conditions, the diffusion coefficient D used in the reconstruction method was the same for the four sets of displacement data and was calculated by means of the Blanc's law [15] to be 1.44×10^{-3} m^2/s for the data at 1.6 kPa and 2.47×10^{-3} m^2/s for the data at 920 Pa. Once again, the reconstruction method failed in extracting the correct velocity profile from the displacement data. The same issue encountered in the velocity reconstruction from the displacement data of argon–acetone flow at 6.5 kPa (Figure 16) appeared here in an even more dramatic way. The reconstructed slip velocity ranged between −16 and +20 m/s, while the expected analytical values should be around 2 m/s, when Maxwell boundary conditions were employed, and full accommodation of the momentum was assumed. The velocity profiles that result from the application of the reconstruction method to the displacement data of Figure 19 were not here presented because qualitatively identical to those reported in Figure 16b, while differing only for the higher fluctuations on the slip velocity.

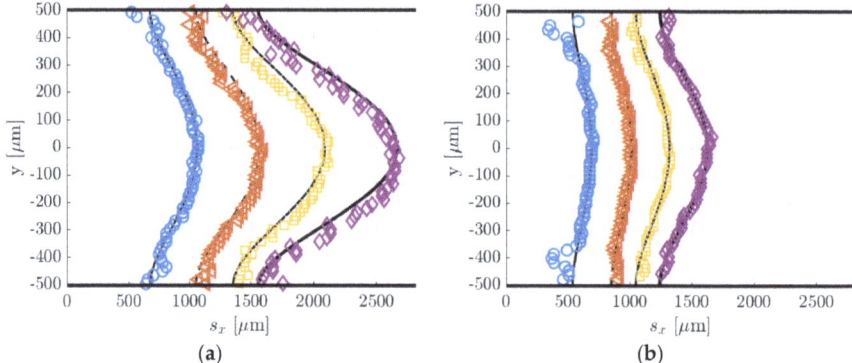

Figure 19. (a) Displacement data computed by applying the Gaussian fitting per line to the MTV images at (a) $p_m = 1580$ Pa (Figure 17) and (b) $p_m = 920$ Pa (Figure 18): $\Delta t = 20$ μs (O), 30 μs (◁), 40 μs (□), and 50 μs (◇). The black lines represent the numerical solutions for the displacement profile provided by the reconstruction method. The two figures are on the same spatial scale.

For a given set of parameters, the reconstruction method provides the velocity profile that generates the molecular displacement that best fits the displacement data points. For this reason, despite the errors on the velocity profiles, the reconstructed displacement profiles (black lines in Figure 18a) always followed very well the displacement data. The inaccuracies on the reconstructed local velocity were likely due to inaccuracies on the height H and diffusion coefficient D used in the reconstruction algorithm (see [13] for more details on the reconstruction method). The reconstructed method is quite sensitive to these parameters, and its sensitivity increases as the molecular slip at the wall gains importance. For the argon–acetone flow at 28 kPa, the sensitivity of the reconstruction to H and D was much lower because of the limited molecular slip at the wall by Taylor dispersion. For this reason, although the inaccuracy on H and D was the same for every experiment, the reconstruction method provided more accurate velocity profiles at high pressures.

Despite the inefficacy of the current reconstruction algorithm in extracting accurate information on the local velocity, the sets of displacement data reported in Figure 19a,b appeared to be qualitatively very good, in the sense that the data dispersion around the reconstructed displacement profiles was very limited. Especially for the data of Figure 19b, despite the stronger Taylor dispersion that increased the molecular slip at the wall and flattened the displacement profile, the current MTV setup was still able to capture accurate information on the gas molecular displacement even in gas flows at pressures as low as 920 Pa. Relatively large fluctuations of the displacement data were only encountered for the case at 920 Pa and at $\Delta t = 20$ μs in regions close to the walls. This is because the corresponding MTV image (Figure 18a) was somehow characterized by a quite low SNR in those regions, where the Gaussian fitting per line failed to identify the correct position of the tagged line.

The displacement profiles of Figure 19a,b were represented on the same spatial scale to allow an easier comparison between the two cases. It was evident how the displacement profiles at 920 Pa had a much lower thickness, which is defined as $\Delta s_x(t) = \max(s_x(y)) - \min(s_x(y))$ [15], as a result of the increased Taylor dispersion. In comparison to the displacement profiles related to argon–acetone flows of Figures 14a and 16a, the displacement profiles in Figure 18a had a larger thickness. Even though the higher molecular diffusion tended to make the displacement profile flatter, the helium flow at 1.6 kPa was characterized by a speed that was about 4 times higher than the argon flow speed, thus making the molecular advection more pronounced. However, for the helium–acetone flow at 920 Pa, the molecular diffusion was so strong that, even though its flow speed was about 3 times that of the argon–acetone flow at 6.5 kPa, the measured displacement thickness resulted in being of the same order of magnitude of that measured in argon flow. Table 5 reports the values of the thickness Δs_x related to the data of helium–acetone flows and the data of argon–acetone flow at 6.5 kPa.

Table 5. Thickness Δs_x of the displacement profile related to the data of argon–acetone flow at 6 kPa (Figure 16a) and of helium–acetone flows at 1.6 kPa (Figure 18a) and at 920 Pa (Figure 18b).

Δt (µs)	Argon–acetone at 6.5 kPa	Helium–acetone at 1.6 kPa	Helium–acetone at 920 Pa
20	206 µm	384.7 µm	149.3 µm
30	298.3 µm	528.5 µm	159.3 µm
40	334.96 µm	748.3 µm	271.9 µm
50	398.35 µm	1124.98 µm	380.8 µm

6.3. Comparison between MTV and CV Data

In this section, MTV and CV techniques were cross-validated by comparison of the measured average velocity at the test section. Since the average velocity $\bar{u}_{CV}$ is derived from the mass flow rate data through analytical formulas, the following data comparison represents also a validation of the theoretical models of the pressure distribution along the channel length of Arkilic et al. [22] and the velocity distribution across the channel section of Ebert and Sparrow [19]. In order to make the data comparison more meaningful, the results were also provided with estimated uncertainty values. The uncertainty ranges reported in Table 6, Table 7, and Table 8 correspond to ± 2 standard deviations, with an associated confidence interval of 95%.

Table 6. Average velocities $\bar{u}_{CV}$ and $\bar{u}_{MTV}$ with estimated uncertainties, and relative deviation between $\bar{u}_{CV}$ and $\bar{u}_{MTV}$ calculated for the image data of Figure 13. The uncertainty range has a confidence level of 95%.

Δt (µs)	$\bar{u}_{CV}$ (m/s)	$\bar{u}_{MTV}$ (m/s)	$(\bar{u}_{CV}-\bar{u}_{MTV})/\bar{u}_{MTV}$ (%)
10	12.42 ± 0.74	10.47 ± 3	18.6
25	12.48 ± 0.74	11.21 ± 1.2	11.3
50	12.54 ± 0.76	11.93 ± 0.6	5.1

Table 7. Average velocities $\bar{u}_{CV}$ and $\bar{u}_{MTV}$ with estimated uncertainties, and relative deviation between $\bar{u}_{CV}$ and $\bar{u}_{MTV}$ calculated for the image data of Figure 15. The uncertainty range has a confidence level of 95%.

Δt (µs)	$\bar{u}_{CV}$ (m/s)	$\bar{u}_{MTV}$ (m/s)	$(\bar{u}_{CV}-\bar{u}_{MTV})/\bar{u}_{MTV}$ (%)
20	10.16 ± 0.6	9.08 ± 1.52	11.9
30	10.06 ± 0.6	9.83 ± 0.98	2.3
40	9.95 ± 0.6	9.92 ± 0.76	0.3
50	9.84 ± 0.6	9.9 ± 0.6	0.6

Table 8. Average velocities $\bar{u}_{CV}$ and $\bar{u}_{MTV}$ with estimated uncertainties, and relative deviation between $\bar{u}_{CV}$ and $\bar{u}_{MTV}$ calculated for the image data of Figure 17. The uncertainty range has a confidence level of 95%.

Δt (µs)	$\bar{u}_{CV}$ (m/s)	$\bar{u}_{MTV}$ (m/s)	$(\bar{u}_{CV}-\bar{u}_{MTV})/\bar{u}_{MTV}$ (%)
20	41.2 ± 2.48	44.59 ± 1.52	7.6
30	41.35 ± 2.48	44.47 ± 0.98	7
40	41.28 ± 2.48	43.78 ± 0.76	5.7
50	41.15 ± 2.46	43.09 ± 0.6	4.5

For the CV measurements, the uncertainty on $\bar{u}_{CV}$ was determined by the uncertainty on the average velocity $\bar{u}_{CV,2D}$ over the cross-section. Using Equation (2) in Equation (5) yields

$$\bar{u}_{CV,2D} = -\frac{V}{Hb\, p_m} \frac{dp_1}{dt}, \qquad (7)$$

which shows how $\bar{u}_{CV,2D}$ depends neither on the gas temperature T in the upstream reservoir nor on the acetone concentration χ composing the gas mixture flowing through the channel. The main uncertainty was given by the channel height, which has been measured by CCD imaging of the Suprasil® windows re-emission. The uncertainty on H was, therefore, estimated based on the CCD resolution, and, because the employed measurement technique has limited accuracy, was conservatively estimated to be twice a pixel size on the CCD. A relative standard deviation on $\bar{u}_{CV}$ of 3% was computed by Taylor expansion of Equation (7) and by assuming that all parameters of the equation are statistically uncorrelated [27].

For the MTV measurements, the average velocity $\bar{u}_{MTV}$ was computed by averaging the reconstructed velocity profiles over the height. Although the local velocity results provided by the reconstruction method depend on the accuracy of the height H and the diffusion coefficient D used in the algorithm, the average velocity $\bar{u}_{MTV}$ only depends on the measured mean displacement $\bar{s}_x$ and the delay Δt. This explained why, even though the MTV velocity profiles might be inaccurate, the reconstruction method always provided profiles with an average velocity that is identical to an average velocity computed simply as

$$\bar{u}_{MTV} = \frac{\bar{s}_x}{\Delta t}. \qquad (8)$$

As for Equation (7), $\bar{s}_x$ and Δt are assumed to be statistically uncorrelated. The mean displacement $\bar{s}_x$ was calculated from the displacement profiles $s_x(y)$ provided by the reconstruction method. For most of the experimental measurements presented in Sections 6.1 and 6.2, the fluctuations of the displacement data point $s_{x,j}$ around the numerical solution $s_x(y)$ were relatively limited; therefore the computation of $\bar{s}_x$ from the displacement data practically provided the same results. When the data fluctuation is stronger, e.g., as for the data at $\Delta t = 20$ s in Figure 19b, using the reconstructed displacement profile for the computation of $\bar{s}_x$ helps in filtering the outlier data out. The delay Δt was known with an accuracy of the order of 5 ns, that is the precision of the PTU in synchronizing the triggering signals. Thus, for the lowest delays of acquisition employed, the maximum relative uncertainty on Δt was only 0.05%. Therefore, since the uncertainty on Δt was negligible, the accuracy on the average velocity $\bar{u}_{MTV}$ was solely determined by the accuracy on $\bar{s}_x$. Uncertainties on the measured mean displacement were introduced by the limited CCD resolution and by the low SNR of the image. The magnitude of the mean displacement determines whether the uncertainty on $\bar{s}_x$ is dominated by the former or the latter type of errors. The estimation of the uncertainty due to low SNR was, however, a difficult task, and more MTV data were required for its quantification. Therefore, a standard deviation of 1 pixel size, which is 15.2 µm, was associated to the measurement of $\bar{s}_x$ with the awareness of possibly underestimating it.

6.3.1. Experiments in Argon–acetone Flow at 28 kPa

Table 6 reports the average velocities measured by MTV and the CV technique, with the corresponding uncertainty ranges. The relative error between MTV and CV measurements is also reported. Especially at high delays, the comparison demonstrated a good match between $\bar{u}_{CV}$ and $\bar{u}_{MTV}$. The mean displacements at $\Delta t = 10, 25$, and 50 µs were, respectively, $\bar{s}_x = 103.87, 280.97$, and 598.56 µm, and thus the corresponding relative standard deviations on $\bar{u}_{MTV}$ due to the CCD spatial resolution were 14.6%, 5.4%, and 2.5%. The small variations on the measured values of $\bar{u}_{CV}$ followed the variations on the inlet–outlet pressure difference characterizing the average MTV image (see Table 1). At $\Delta t = 10$ µs, the SNR of the image (Figure 14a) was relatively high, but the mean displacement was about 7 pixels only, which explained the relatively high discrepancy with respect to the CV measurement. In this case, the CCD resolution dominated the uncertainty on $\bar{s}_x$, thus the estimated error range on $\bar{u}_{MTV}$ was most likely correct. This was also suggested by the fact that the relative deviation between MTV and CV measurements decreased as a function of the delay. At higher delays, however, the uncertainty range might be underestimated, since the SNR of the image visibly was reduced (Figure 14c).

6.3.2. Experiments in Argon–acetone Flow at 6.5 kPa

Despite the inaccuracy on the reconstructed velocity profile, the average velocity $\bar{u}_{MTV}$ was very close to the values $\bar{u}_{CV}$ calculated from the mass flow rate measurements (Table 7). As for the case at higher pressure, higher delay times provide a better match of the two measurements. The measured mean displacements were $\bar{s}_x$ = 181.4, 294.8, 396.5, and 494.6 µm, and the resulting relative standard deviations on $\bar{u}_{MTV}$ were 8.4%, 5%, 3.8%, and 3%, respectively for Δt = 20, 30, 40, and 50 µs. The Taylor dispersion amplified the molecular slip at the wall, which contributed more to the resulting mean displacement. Therefore, the imprecision in capturing the molecular slip at the wall contributed more to increase the uncertainty on the average velocity. Nevertheless, the SNR of the images of Figure 15 was relatively high, and the measured $\bar{u}_{MTV}$ differed from the values $\bar{u}_{CV}$ by less than 3% at Δt = 30 µs and only by about 0.5% at Δt = 40 and 50 µs. The variations on $\bar{u}_{CV}$ measured at different delays were limited to about 3% and reflected the slight decrease on the pressure difference characterizing the MTV images (see Table 2). The decreasing relative deviation between CV and MTV measurements indicates that the variations on $\bar{u}_{MTV}$ measured at different delays were mainly determined by the inaccuracies on the measured molecular displacement due to the CCD resolution.

6.3.3. Experiments in Helium–acetone Flows

As for the case of argon–acetone flow, even though the reconstructed velocity was inaccurate, the average velocity was well extracted from the displacement data and was always approximately the same, regardless of the delay considered. Tables 8 and 9 compared the values of $\bar{u}_{MTV}$ with those of $\bar{u}_{CV}$. Since the helium–acetone flow was characterized by much higher mean molecular displacements, the relative standard deviation on $\bar{u}_{MTV}$ given by the CCD resolution was at most 1.7%, for the data at 1.6 kPa, and 2.4%, for the data at 920 Pa.

Table 9. Average velocities $\bar{u}_{CV}$ and $\bar{u}_{MTV}$ with estimated uncertainties, and relative deviation between $\bar{u}_{CV}$ and $\bar{u}_{MTV}$ calculated for the image data of Figure 18. The uncertainty range has a confidence level of 95%.

Δt (µs)	$\bar{u}_{CV}$ (m/s)	$\bar{u}_{MTV}$ (m/s)	$(\bar{u}_{CV}-\bar{u}_{MTV})/\bar{u}_{MTV}$ (%)
20	26.72 ± 1.6	31.06 ± 1.52	14
30	26.6 ± 1.6	31.24 ± 0.98	14.8
40	26.7 ± 1.6	29.81 ± 0.76	10.4
50	26.77 ± 1.6	28.89 ± 0.6	7.3

The highest values of $\bar{s}_x$ were recorded for the helium flow at 1.6 kPa, where it varied from 0.89 mm at Δt = 20 µs to 2.15 mm at Δt = 50 µs, which was more than 4 times the displacement imaged in the argon–acetone flows. Even for the helium flow at 920 Pa, the measured mean displacements were relatively high: $\bar{s}_x$ = 620.22, 936.58, 1192.21, and 1444.39 µm, respectively for Δt = 20, 30, 40, and 50 µs. Therefore, the uncertainty on $\bar{u}_{MTV}$ was likely dominated by the SNR of the images. The molecular displacement at the wall was large and comparable to the centerline displacement, and therefore it contributed considerably to the measured $\bar{s}_x$. Since the SNR was lower in regions close to the channel walls, inaccuracies in measuring the displacement slip at the wall might bias the measured mean displacement. Moreover, the high molecular diffusion makes the tagged line thicker, which reduces the accuracy of the Gaussian fitting procedure in localizing the peak position. These considerations suggest that the uncertainties for $\bar{u}_{MTV}$ reported in Tables 8 and 9 were probably underestimated. For the case at 920 Pa, the molecular slip at the wall represented between 85% and 90% of the mean molecular displacement. This might explain why the difference between CV and MTV measurements were higher than for the case at 1.6 kPa, where the molecular slip at the wall contributed to the mean displacement at about 75%. However, the 4 values of $\bar{u}_{MTV}$ were measured by repeating the experiment 4 different times, and the MTV velocity measured at different delays varied around the average value of 2% for the data at 1.6 kPa and 4.5% for the data at 920 Pa. These variations were of the same order of magnitude of the estimated variations due to fluctuations of the thermodynamic conditions characterizing the 20 averaged images (Section 6.2). As it can be noted from Tables 8 and 9, the value of $\bar{u}_{MTV}$ was always higher than that of $\bar{u}_{CV}$, and their relative deviation was on average about 6.2% for the flow at 1.6 kPa and 11.6% for the flow at 920 Pa. This suggests the existence of a constant biasing error affecting the measurements. While MTV measurements are only based on the optical tracking of the gas displacement, the calculation of the CV average velocity makes use of the measured dimensions of the channel cross-section. At working pressures of the order of 1 kPa, the distance between the Suprasil® windows might be slightly reduced by the higher external pressure. Therefore, the bias between CV and MTV average velocities was probably partly due to the uncertainty on the value of H. The idea that the compression of the channel height should be higher at lower pressures was reflected by the higher differences between CV and MTV data obtained for the case at 920 Pa. Variations of 6.2% and 11% on the channel height could correspond to a displacement towards the interior of the channel of only 30 µm and 55 µm, respectively, for each Suprasil® window. In addition, if the windows were no longer perfectly aligned with the channel walls, this misalignment could create local velocity perturbation that affect the measured displacement profile.

Since the images of Figures 17 and 18 represent the very same flow conditions and, thus, the same velocity profile, the mean displacement of the tagged lines evolved linearly with time, as demonstrated in Figure 20.

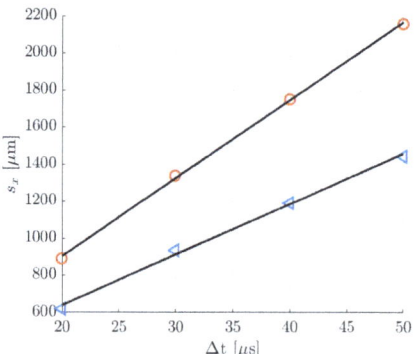

Figure 20. Mean molecular displacement $\bar{s}_x$ in helium flow at delays Δt = 20, 30, 40, and 50 µs: (O) data at 1.6 kPa; (◁) data at 920 Pa. The black solid line represents the linear interpolation of the MTV data, which provides an average velocity $\bar{u}_{MTV}$ = 42.2 m/s for the data at 1.6 kPa and $\bar{u}_{MTV}$ = 27.28 m/s for the data at 920 Pa.

A linear interpolation of these data provided a second measurement for the average velocity. For the case at 1.6 kPa, this second measurement was $\bar{u}'_{MTV}$ = 42.2 m/s, which differed by about 4% from the average of the reconstructed velocity profiles, and only 2.3% from the average velocity measured by the CV technique. For the flow at 920 Pa, $\bar{u}'_{MTV}$ = 27.28 m/s, which was smaller than the average of the reconstructed velocity profiles by 10% but differed only by 2.2% from the CV value. The MTV velocity that resulted from the linear fitting was likely more accurate than the average velocities measured from each MTV image individually, since more displacement data were used at once. This final result suggests the idea that the reconstruction algorithm might provide more accurate velocity profiles if properly modified to allow processing all the displacement data at different delays simultaneously.

7. Discussions and Conclusions

This work demonstrated a successful application of the MTV technique to low-pressure gas flows in a millimetric channel. To the best of our knowledge, the images of Figures 17 and 18 are the first flow visualizations ever reported of a gas flow in the slip regime and in a confined domain. MTV data were reported for gas flows with Knudsen numbers as high as $Kn = 0.014$. For the case of helium–acetone flow, a slip velocity at the wall of about 2 m/s is expected when a Maxwell boundary condition is used, which indicates that rarefaction has significant effects at the wall in the investigated flow conditions. As shown in Figures 14a, 16a and 18a,b, MTV could provide, even at pressures as low as 920 Pa, accurate data on the gas displacement, which was very well fitted by the numerical prediction of the 1D advection–diffusion equation. In particular, the MTV data here reported gave clear experimental evidences of the strong effects produced by the Taylor dispersion on the molecular displacement, which was, up to now, only theoretically and numerically predicted [13,15]. The comparison with the CV technique demonstrated also the good accuracy of the MTV technique in providing average displacement and average velocity measurements in rarefied gas flows. Some estimations of the uncertainties on the measured velocities were provided. A better quantification of the errors on the MTV average velocity would require the collection of more experimental data, thus allowing a more complete statistical analysis of the fluctuations on the signal generated by the molecular diffusion, the phosphorescence emission, and the intensified CCD itself. However, as demonstrated in Section 6.2, repeating the same experiments several times and using different delays of acquisition provided substantially the same MTV average velocity, which strongly supported the conclusion that the MTV measurements of the gas molecular displacement were characterized by a good precision, even for the highest Knudsen numbers investigated.

The success of our MTV implementation to the case of low-pressure gas flows in a confined domain was the result of two important experimental achievements. Firstly, an intense phosphorescence signal at pressures of the order of 1 kPa could be obtained only by selecting the optimal wavelength for the acetone excitation, which was 310 nm. In our previous work [15], we demonstrated that this laser wavelength allowed increasing the phosphorescence signal of at least 10 times with respect to an excitation at 266 nm. The second key aspect that made MTV application in confined and low-pressure gas flows possible is the open loop configuration chosen for the gas flow system. Actually, an open loop gas circuit is, for the current version of the tested channel, the only practical option for having an oxygen-free gas mixture at the test section. The tested channel used for MTV measurements needs to guarantee optical accesses for both laser excitation and CCD acquisition. These requirements make the channel be an assembly of different materials, which easily introduces leakages when used in a low-pressure system. This can generate a small flow of oxygen molecules, especially in correspondence of the Suprasil® windows, which tends to quench the tagged acetone molecules. The current open loop configuration overcomes this issue by feeding the test section with fresh acetone–gas mixture at relatively high flow rates. Future improvements of this work envisage a better design of the test channel that aims to make it more leakage-proof. This will further improve the SNR of MTV images and might allow MTV applications even in a closed loop gas circuit, which has the advantage over the open loop version of providing more stable thermodynamic conditions for MTV acquisitions.

In non-rarefied flows and for pressure conditions at which the effects of the molecular diffusion on the evolution of the molecular displacement profile were limited, the reconstruction method could provide accurate local velocity measurements along the channel height and could correctly predict a negligible slip velocity at the wall. However, when the gas pressure was low enough, the Taylor dispersion significantly distorted the tagged line, and thus the current version of the reconstruction method was unable to extract accurate information on the local gas velocity, despite the very good match between MTV and CV measurements and the relatively high quality of the MTV images and of the measured displacement profiles. Further investigation is necessary to understand whether the available MTV dataset is not large enough or the post-processing method needs to be improved (or both).

Preliminary analysis revealed a significant sensitivity of the reconstructed velocity profile to the parameters H and D. Future efforts aim to better characterize the sensitivity of the post-processing algorithm and to improve the accuracy on the measurements of the channel height and of the diffusion coefficient. An important challenge related to the latter intent is given by the fact that the diffusion coefficient D strongly depends on the value used for the acetone molecular diameter d. In this work, the value $d = 470$ pm provided by Van der Perre et al. [24] was considered because this result is supported by several other authors [25,26]. Nevertheless, there are also other works in the literature that provide quite different estimations of the acetone kinetic diameter, which allow variations from 460 to 730 pm (590 pm by Almy et al. [28], 616 pm in Nydatok et al. [29], and 730 pm in Frezzotti et al. [13]). These uncertainties on the molecular diameter introduce relative uncertainties on the diffusion coefficient D that can be as large as 100%. New experimental data on d and D would be, therefore, of great interest for the improvement of this work. Differently from the parameter D, channel height H can be directly measured, thus the uncertainty on this parameter can be highly reduced. Currently, the accuracy on H is mainly affected by the difficulty in controlling the Suprasil® windows position, which can slightly change as a function of the working gas pressure. Future new designs of the channel test section will aim to solve this issue, thus drastically improving the accuracy on the measurement of H. Nevertheless, since the displacement profile solution of the advection–diffusion equation depends on H^2 [13], the reconstruction method is much more sensitive to H than to D. In this context, new versions of the velocity reconstruction algorithm that are more robust to uncertainties on the parameters H and D are under development.

Furthermore, it is also plausible that the advection–diffusion equation used for the velocity reconstruction does not capture all the physical features of the pressure-driven binary gas flow. For instance, the high molecular mass ratio of the helium–acetone mixture and the relatively strong pressure gradient driving the investigated gas flow might generate a velocity difference between the two species, thus leading to an effect known as gas separation [30]. Future efforts will also aim to investigate (i) if this phenomenon introduces a non-negligible velocity bias between helium and acetone molecules and (ii) if it is a key mechanism that needs to be embedded in the reconstruction algorithm.

In conclusion, despite the new challenges encountered in carrying out local velocity measurements in gas flows characterized by high molecular diffusion, this work demonstrated that MTV is currently the most promising technique for providing direct measurements of the slip velocity at the wall in rarefied gas flows.

Author Contributions: Conceptualization, D.F., M.R.-C., C.B. and S.C.; methodology, D.F., M.R.-C., C.B. and S.C.; software, D.F.; validation, D.F., M.R.-C., C.B. and S.C.; formal analysis, D.F.; investigation, D.F.; resources, L.B. and S.C.; data curation, D.F.; writing—original draft preparation, D.F. and M.R.-C.; writing—review and editing, D.F., M.R.-C., C.B., L.B. and S.C.; visualization, D.F.; supervision, M.R.-C., C.B., L.B. and S.C.; project administration, L.B. and S.C.; funding acquisition, L.B. and S.C. All authors have read and agreed to the published version of the manuscript. All the reported work, except a few data treatments and the writing of the paper, have been performed at Institut Clément Ader (ICA), Toulouse.

Funding: This research was funded by the European Community's Seventh Framework Program (FP7/2007-2013) under grant agreement n 215504, and from Féderation de Recherche Fermat, FR 3089.

Conflicts of Interest: The authors declare no conflict of interest.

References

1. Gomez, J.; Groll, R. Pressure drop and thrust predictions for transonic micronozzle flows. *Phys. Fluids* **2016**, *28*, 022008. [CrossRef]
2. Cattafesta, L.N.; Sheplak, M. Actuators for active flow control. *Annu. Rev. Fluid Mech.* **2011**, *43*, 247–272. [CrossRef]

3. Gerken, I.; Brandner, J.J.; Dittmeyer, R. Heat transfer enhancement with gas-to-gas micro heat exchangers. *Appl. Eng.* **2016**, *93*, 1410–1416. [CrossRef]
4. An, S.; Gupta, N.K.; Gianchandani, Y.B. A Si-micromachined 162-stage two-part Knudsen pump for on-chip vacuum. *J. Microelectromech. Syst.* **2014**, *23*, 406–416. [CrossRef]
5. Colin, S. Rarefaction and compressibility effects on steady and transient gas flows in microchannels. *Microfluid. Nanofluid.* **2005**, *1*, 268–279. [CrossRef]
6. Pitakarnnop, J.; Varoutis, S.; Valougeorgis, D.; Geoffroy, S.; Baldas, L.; Colin, S. A novel experimental setup for gas microflows. *Microfluid. Nanofluid.* **2010**, *8*, 57–72. [CrossRef]
7. Silva, E.; Rojas-Cardenas, M.; Deschamps, C.J. Experimental analysis of velocity slip at the wall for gas flows of nitrogen, R134a, and R600a through a metallic microtube. *Int. J. Refrig.* **2016**, *66*, 121–132. [CrossRef]
8. Miles, R.; Cohen, C.; Connors, J.; Howard, P.; Huang, S.; Markovitz, E.; Russell, G. Velocity measurements by vibrational tagging and fluorescent probing of oxygen. *Opt. Lett.* **1987**, *12*, 861–863. [CrossRef]
9. Pitz, R.W.; Brown, T.M.; Nandula, S.P.; Skaggs, P.A.; DeBarber, P.A.; Brown, M.S.; Segall, J. Unseeded velocity measurement by ozone tagging velocimetry. *Opt. Lett.* **1996**, *21*, 755–757. [CrossRef]
10. Stier, B.; Koochesfahani, M.M. Molecular tagging velocimetry (MTV) measurements in gas phase flows. *Exp. Fluids* **1999**, *26*, 297–304. [CrossRef]
11. Lempert, W.R.; Jiang, N.; Sethuram, S.; Samimy, M. Molecular tagging velocimetry measurements in supersonic microjets. *AIAA J.* **2002**, *40*, 1065–1070. [CrossRef]
12. Samouda, F.; Colin, S.; Barrot, C.; Baldas, L.; Brandner, J.J. Micro molecular tagging velocimetry for analysis of gas flows in mini and micro systems. *Microsyst. Technol.* **2015**, *21*, 527–537. [CrossRef]
13. Frezzotti, A.; Si Hadj Mohand, H.; Barrot, C.; Colin, S. Role of diffusion on molecular tagging velocimetry technique for rarefied gas flow analysis. *Microfluid. Nanofluid.* **2015**, *19*, 1335–1348. [CrossRef]
14. Si Hadj Mohand, H.; Frezzotti, A.; Brandner, J.J.; Barrot, C.; Colin, S. Molecular tagging velocimetry by direct phosphorescence in gas microflows: Correction of Taylor dispersion. *Exp. Fluid Sci.* **2017**, *83*, 177–190. [CrossRef]
15. Fratantonio, D.; Rojas-Cardenas, M.; Si Hadj Mohand, H.; Barrot, C.; Baldas, L.; Colin, S. Molecular tagging velocimetry for confined rarefied gas flows: Phosphorescence emission measurements at low pressure. *Exp. Fluid Sci.* **2018**, *99*, 510–524. [CrossRef]
16. Lozano, A.; Yip, B.; Hanson, R.K. Acetone: A tracer for concentration measurements in gaseous flows by planar laser-induced fluorescence. *Exp. Fluids* **1992**, *13*, 369–376. [CrossRef]
17. Costela, A.; Crespo, M.T.; Figuera, J.M. Laser photolysis of acetone at 308 nm. *J. Photochem.* **1986**, *34*, 165–173. [CrossRef]
18. Pringsheim, P. *Fluorescence and Phosphorescence*; Interscience Publishers: New York, NY, USA, 1949.
19. Ebert, W.A.; Sparrow, E.M. Slip flow in rectangular and annular ducts. *J. Basic Eng.* **1965**, *87*, 1018–1024. [CrossRef]
20. Lempert, W.R.; Boehm, M.; Jiang, N.; Gimelshein, S.; Levin, D. Comparison of molecular tagging velocimetry data and direct simulation Monte Carlo simulations in supersonic micro jet flows. *Exp. Fluids* **2003**, *34*, 403–411. [CrossRef]
21. Silva, E.; Deschamps, C.J.; Rojas-Cardenas, M.; Barrot, C.; Baldas, L.; Colin, S. A time-dependent method for the measurement of mass flow rate of gases in microchannels. *Int. J. Heat Mass Transf.* **2018**, *120*, 422–434. [CrossRef]
22. Arkilic, E.B.; Schmidt, M.A.; Breuer, K.S. Gaseous slip flow in long microchannels. *J. Microelectromech. Syst.* **1997**, *6*, 167–178. [CrossRef]
23. Harley, J.C.; Huang, Y.; Bau, H.H.; Zemel, J.N. Gas flow in micro-channels. *J. Fluid Mech.* **1995**, *284*, 257–274. [CrossRef]
24. Van der Perre, S.; Van Assche, T.; Bozbiyikm, B.; Lannoeye, J.; De Vos, D.E.; Baron, G.V.; Denayer, J.F. Adsorptive characterization of the ZIF-68 metal-organic framework: A complex structure with amphiphilic properties. *Langmuir* **2014**, *30*, 8416–8424. [CrossRef] [PubMed]
25. Song, Z.; Huang, Y.; Xu, W.L.; Wang, L.; Bao, Y.; Li, S.; Yu, M. Continuously adjustable, molecular-sieving "gate" on 5A zeolite for distinguishing small organic molecules by size. *Sci. Rep.* **2015**, *5*, 13981. [CrossRef] [PubMed]

26. Ganesh, R.S.; Navaneethan, M.; Mani, G.K.; Ponnusamy, S.; Tsuchiya, K.; Muthamizhchelvan, C.; Kawasaki, S.; Hayakawa, Y. Influence of Al doping on the structural, morphological, optical, and gas sensing properties of ZnO nanorods. *J. Alloy. Compd.* **2017**, *698*, 555–564. [CrossRef]
27. BIPM; IEC; IFCC; ILAC; IUPAC; IUPAP; ISO; OIML. *Evaluation of Measurement Data—Guide for the Expression of Uncertainty in Measurement*; JCGM 100:2008; BIPM: Sèvres, France, 2008.
28. Almy, G.M.; Anderson, S. Lifetime of fluorescence in diacetyl and acetone. *J. Chem. Phys.* **1940**, *8*, 805–814. [CrossRef]
29. Nadykto, A.B.; Yu, F. Uptake of neutral polar vapor molecules by charged clusters/particles: Enhancement due to dipole-charge interaction. *J. Geophys.* **2003**, *108*, 4717. [CrossRef]
30. Szalmas, L.; Pitakarnnop, J.; Geoffroy, S.; Colin, S.; Valougeorgis, D. Comparative study between computational and experimental results for binary rarefied gas flows through long microchannels. *Microfluid. Nanofluid.* **2010**, *9*, 1103–1114. [CrossRef]

© 2020 by the authors. Licensee MDPI, Basel, Switzerland. This article is an open access article distributed under the terms and conditions of the Creative Commons Attribution (CC BY) license (http://creativecommons.org/licenses/by/4.0/).

Article

Measurement of Heat Transfer from Anodic Oxide Film on Aluminum in High Knudsen Number Flows

Hiroki Yamaguchi * and Kenji Kito

Department of Micro-Nano Mechanical Science and Engineering, Nagoya University, Nagoya 464-8603, Japan; kito.kenji@a.mbox.nagoya-u.ac.jp
* Correspondence: hiroki@nagoya-u.jp; Tel.: +81-52-789-2702

Received: 24 January 2020; Accepted: 24 February 2020; Published: 25 February 2020

Abstract: The heat transfer in vacuum depends on the gas–surface interaction. In this study, the heat flux from anodic oxide films on aluminum with different anodizing times through a gas confined between two surfaces with different temperatures was studied. We prepared a non-treated surface, a surface with a normal anodizing time of 30 min, and a surface with 90 min, where the formed film would partially dissolve by long time exposure to the solution. The formation of the films was checked by electrical resistance. Scanning electron microscope (SEM) images were obtained for the three sample surfaces. Even though it was difficult to observe the hexagonal cylindrical cell structures on anodic oxide films, the 30 min sample surface was shown to be rough, and it was relatively smooth and powdery for the 90 min sample surface. The heat fluxes from three sample surfaces were measured from the free-molecular to near free-molecular flow regimes, and analyzed to obtain the energy accommodation coefficients. The heat fluxes were well fitted by the fitting curves. The energy accommodation coefficients for both helium and argon increased by anodizing an aluminum sample surface, while they decreased with increasing the anodizing time up to 90 min indicating the dissolution of the film.

Keywords: gas–surface interaction; thermal accommodation coefficient; vacuum

---

1. Introduction

The heat transfer from a hot to a cold surface in vacuum is a basic problem. In vacuum, the Knudsen number is large due to a large mean free path of gas molecules. In such a high Knudsen number flow, the number of collisions between gas molecules and a solid surface cannot be neglected compared with that between gas molecules. Then, the gas–surface interaction plays an important role in the heat transfer problem.

The gas–surface interaction, which is a scattering process of gas molecules from a solid surface at the boundary of a thermal-fluid field, is known to be a complicated process depending on many parameters of gas species and a solid surface [1]. As the boundary condition of a thermal-fluid field, the statistical behavior of molecules is important and convenient for analysis. For such purpose in the gas–surface interaction, the accommodation coefficient [2], which is an integral characteristics of the interaction, is often employed in models for the gas–surface interaction. The accommodation coefficient represents mean transfer rate, probability, fraction or efficiency of exchanging physical properties between gas molecules and a solid surface through the interaction. For the heat transfer problem, energy transfer is related; thus, we focus on the energy accommodation coefficient (EAC) or the thermal accommodation coefficient, which are equivalent for a static equilibrium gas. The EAC α is defined as [1–3]

$$\alpha = \frac{E_i - E_r}{E_i - E_s}, \tag{1}$$

where E_i, E_r and E_s are the mean incident and reflected energy fluxes, and the energy flux of gas molecules fully accommodated to the surface, respectively.

The EACs or the thermal accommodation coefficients for various pairs of gas species and surface materials have been measured for a long time [2]. Although there are quite large scatterings in the measured values for gas–surface pairs, several qualitative characteristics are discussed. For the effect of the surface roughness, it is known to have a large EAC for a rough surface, i.e., gas molecules accommodate well to a rough surface, because of multiple collisions [1]. Therefore, engineering surfaces have been considered to have the EAC around unity. Recent study [4] showed the effect of the surface roughness on the EAC by comparing the results on the machined, polished and deposited surfaces with or without plasma treatment, showing that the surface roughness appeared to have only a minor effect. The rms surface roughness of these surfaces was reported as ~2 µm for the machined surfaces and ~0.02 µm for the polished surfaces.

In this study, the effect of surface roughness on the energy accommodation coefficient is studied by employing an anodic oxide film on aluminum. An anodic oxide film has the hexagonal cylindrical cell structure with several to several hundreds nm diameter [5,6]. The structure with many pores of such size, similar to porous materials, may cause multiple collisions of gas molecules. Therefore, it is suited to see the effect of roughness on the EAC. It is also important to note that the anodization process is a wet process; thus, the process would roughen all the area of sample surfaces even though there are distortions, large adsorbates or dimples on sample surfaces.

2. Materials and Methods

The heat flux from a sample surface of an anodic oxide film on aluminum to a cold vacuum chamber in vacuum as a function of pressure was measured to extract the EAC.

2.1. Sample Surfaces

The sample surfaces of the anodic oxide film were prepared by anodizing a 0.3-mm-thick aluminum plate (A1050P, AS-ONE, Osaka, Japan), which is a general purpose product. A strip was cut from the plate. Then, it was wiped by acetone, dipped in a solution of NaOH, and rinsed in distilled water. The strip was then immersed in a diluted solution of H_2SO_4 (1 mol/L), and the electric current was applied with the formation voltage of 20 V and the current density of 12.5 mA/cm^2. Chilled water was circulated through a coil placed in the solution to keep its temperature constant at around 5 °C. The anodization time is usually set to 10–30 min for the above conditions [5,6]. It is known that with a long anodization time a hexagonal cylindrical cell structure of an anodized film is chemically dissolved by long time exposure to the solution. To see the effect of the cell structure, we selected 30 min for a normal sample and 90 min for less roughened surface with the anodized material. These two sample surfaces were compared with a non-treated sample surface, which is hereafter called the 0 min sample surface.

First, electrical resistance of the sample surfaces was measured by a tester to check the formation of anodized films on the sample surfaces, since an anodic oxide film is an insulator. It was easily verified that anodized films were formed on both the 30 min and 90 min samples.

To check the surface roughness in detail, the sample surfaces were measured by SEM (JSM-7000F, JEOL, Tokyo, Japan). The obtained scanning electron microscope (SEM) images are shown in Figure 1. We tried to obtain high magnification images; however, it was difficult to observe nano-scale cell structures of the anodic oxide film because of the nature of the film. From these images, it is easily observed that the sample surfaces are quite rough. There are many scratches and roughness already in the 0 min sample surface. Several quite large bumps or dents (black area) appeared on the 30 min sample surface, while small scratches are relatively smoothened. On the other hand, the 90 min sample surface becomes relatively smoother than the 30 min sample surface, but powdery. This might be the result of the dissolution of the film from the top by long time exposure to the solution [5,6] as mentioned above. It is also interesting to see that there are fewer bumps or dents that are observed in the 30 min sample surface.

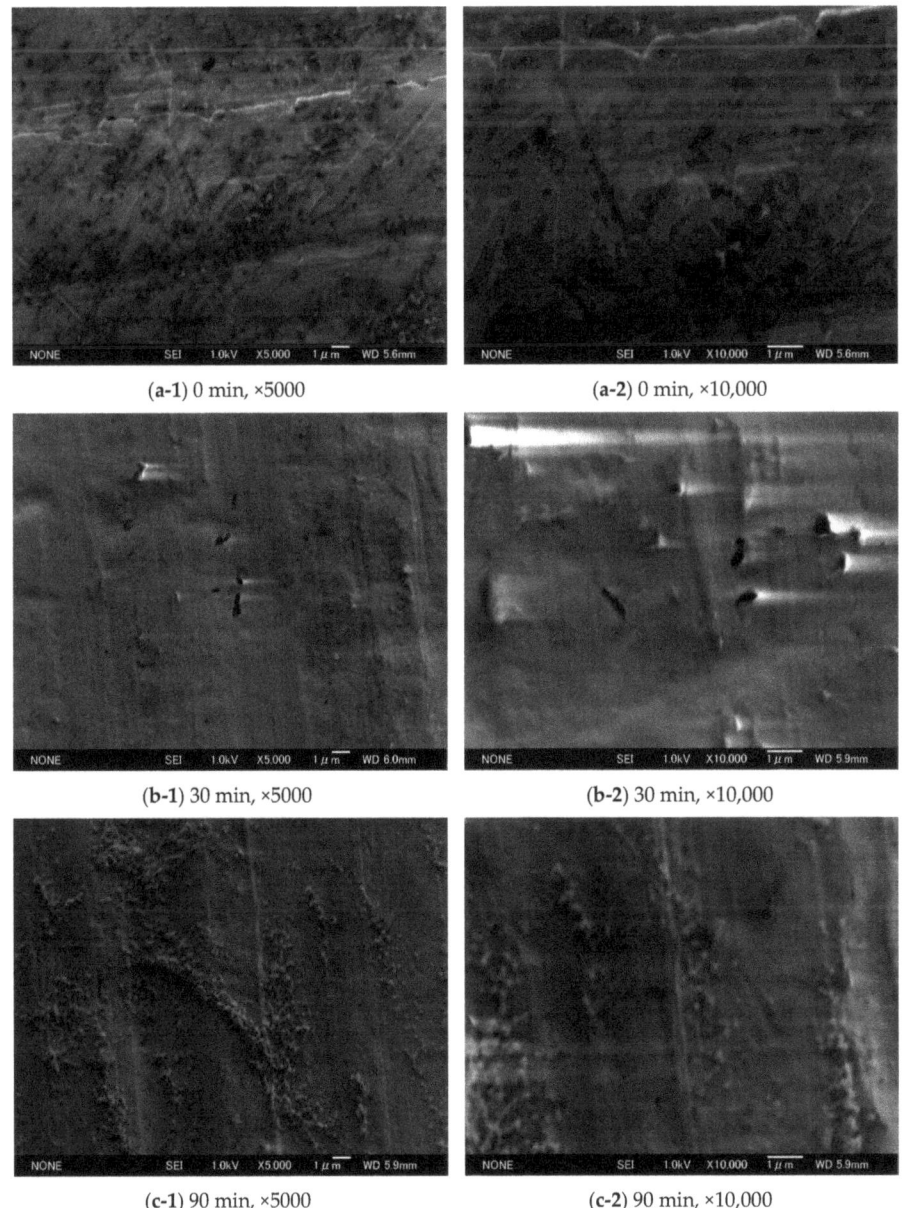

Figure 1. Scanning electron microscope (SEM) images of the anodic oxide films on the sample surfaces: (**a**) 0 min; (**b**) 30 min; and (**c**) 90 min with the magnification of (**1**) ×5000; and (**2**) ×10,000.

2.2. Method

In the free-molecular flow regime, the heat flux between two surfaces with different temperatures is explained by the energy transfer by molecular motions. The heat flux in the free-molecular flow regime q_{FM} is theoretically expressed as

$$q_{FM} = \frac{1}{8}\alpha \frac{\gamma+1}{\gamma-1}\frac{\bar{v}}{T}p\Delta T, \quad \bar{v} = \sqrt{\frac{8kT}{\pi m}}, \quad (2)$$

where γ, $\bar{v}$, T, p, ΔT, k and m are the specific heat ratio, the mean molecular speed, temperature, pressure, the temperature difference of two surfaces, the Boltzmann constant, and the molecular mass of the gas, respectively. The heat flux is proportional to pressure and the EAC; thus, by measuring the heat flux as a function of pressure, the EAC can be derived.

However, as mentioned in our previous studies [7–9], it was not easy to accurately measure only in the free-molecular flow regime, i.e., at a high vacuum condition, in a simplified low-cost apparatus because of small heat flux. Therefore, we used pressure conditions slightly higher than the upper limit of the free-molecular flow regime. A general model expression to describe the heat flux from the free-molecular flow regime up to the continuum flow regime was employed [10,11], which is expressed as

$$\frac{1}{q} = \frac{1}{q_{FM}} + \frac{1}{q_C}, \quad (3)$$

where q_C is the heat flux in the continuum flow regime. The heat flux as a function of pressure would be slightly curved by this expression. The obtained heat flux was fitted by this expression to obtain the EAC in Equation (2).

2.3. Setup

The experimental setup is explained in detail elsewhere [7–9]. A schematic of the experimental set up is shown in Figure 2. A spherical vacuum chamber made by Pyrex, which had a similar shape to a spherical flask, was employed. The inner radius of the chamber R_C was 49.5 mm. The chamber was immersed in a water bath to keep the temperature of the chamber T_C constant. The measured temperature of the chamber surface T_C was about 290 K. The chamber was equipped with NW16 flanges for connections without leakage. The test gas was supplied from commercially available gas cylinders of pure helium and argon. The pressure in the chamber was measured by a temperature-controlled capacitance manometer (Baratron® 627B, MKS Instruments, Andover, MA, USA). The chosen pressure conditions were limited to below 1.4 Pa to be in the near free-molecular flow regime, so that the effect of the general model expression of Equation (3) was minimized.

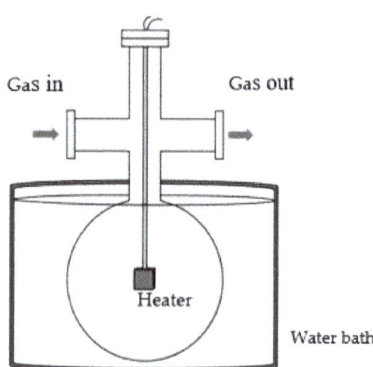

Figure 2. A schematic of the experimental setup.

The sample surfaces of the anodic oxide films were placed on both sides of a tiny flat-shaped heater with the size of 11.8 × 12.0 × 0.38 mm^3 (Toyo Precision Parts MFG, Nara, Japan). The heater with the sample surfaces was placed at the center of the spherical vacuum chamber. Since the heater was small compared with the spherical vacuum chamber, the temperature in Equation (2) could be approximated by the temperature of the chamber T_C due to the large surface area ratio [7–9]. It also appeared that the system could be approximated as a concentric spherical shells system for estimating the heat flux in the continuum limit in Equation (3) [7–9]. The heat flux in the continuum limit q_C was calculated by approximating the sample surfaces as a sphere with radius R_H having the same surface area, and the expression becomes as

$$q_C = \kappa(T_C) \frac{\mathcal{T}^{\omega+1} - 1}{(\omega+1)(\mathcal{T}-1)} \Delta T \frac{R_C R_H}{R_C - R_H} \frac{1}{R_H^2}, \tag{4}$$

where $\kappa(T)$, ω, $\mathcal{T}$ are the thermal conductivity of the gas at temperature T, the thermal conductivity index, and the temperature ratio of two surfaces T_H/T_C, respectively. The thermal conductivity was assumed to be proportional to T^ω following the model with the inverse power law potential for a monatomic gas. On the other hand, the modified expression, which slightly modifies the heat flux in the continuum in Equation (3), was obtained for better fitting to the results of the S-model solutions as [8]

$$\frac{1}{q} = \frac{1}{q_{FM}} + \frac{1}{\zeta q_C}, \quad \zeta = \frac{1}{1 - \frac{c_1}{\delta_0 + c_2}}, \tag{5}$$

where $\delta_0 = \frac{R_C - R_H}{l}$, $l = \frac{\mu(T_C) \bar{v}}{p}$, which is the equivalent mean free path, μ is the viscosity, $c_1 = 1.04\alpha \frac{T_H}{T_C} \frac{R_H}{R_C}$, $c_2 = 1.97\alpha \frac{T_H}{T_C} \frac{R_H}{R_C}$, respectively. We also tried to employ this expression. It was suggested that Equation (5) gave a smaller EAC than Equation (3). The thermal conductivity and viscosity were obtained from [9,12].

An analog electrical bridge circuit was employed to maintain the temperature of the heater by keeping the resistance of the heater element printed by platinum paste constant. The heat transfer rate from the sample surfaces was measured by an electrical consumption to keep the heater temperature constant. The energy consumption consisted of the heat conduction through gas, which we wanted to measure, the radiation and the heat loss through the electrical leads of the heater. Since only the first term depended on pressure, this term could be extracted by evaluating the latter two terms by extrapolating the heat flux to the vacuum limit using values below 0.1 Pa. The convection was negligible due to the low pressure condition. The heat flux q was calculated from the heat transfer rate and the surface area of the samples. The temperature of the sample surfaces was estimated from the electrical resistance of the heater. The calibration curve between temperature and the electrical resistance of the heater with sample surfaces was measured beforehand. The sample surface temperature T_H was about 360 K.

The uncertainty of the measurement was quite difficult to evaluate; however, the error of the measurement was known to be less than 5% [7–9]. Therefore, it was possible to compare the results at least qualitatively between the conditions.

3. Results and Discussions

3.1. Heat Flux

The heat flux as a function of pressure was measured four times for each condition to check the repeatability. Typical examples of the measured heat fluxes and the fitted curves by Equation (3) and Equation (5) for the 0 min, 30 min and 90 min sample surfaces are plotted in Figure 3 for helium and argon. From the figure, the experimentally measured data are well explained by the fitting curves.

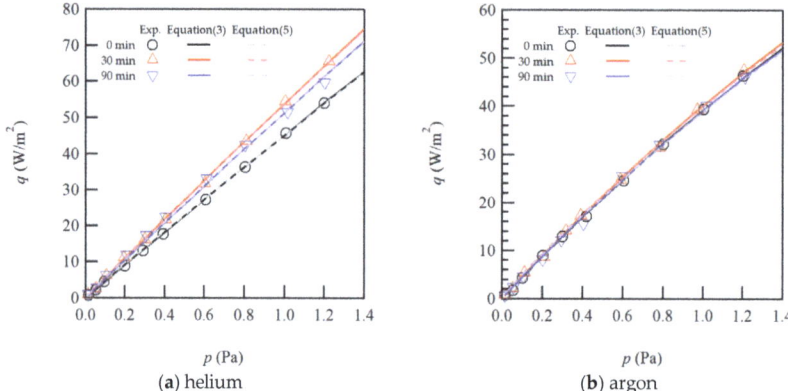

Figure 3. Typical results of the measured heat fluxes and the fitted curves as a function of pressure for 0 min, 30 min and 90 min sample surfaces for (**a**) helium and (**b**) argon.

For the three surface samples, the sizes of the heat flux are clearly different for helium; while they are almost similar for argon. It is well known that helium is quite sensitive to the surface conditions due to its light molecular mass and small size. From the figure for helium, the gradient of the heat flux increases by the anodization. However, it slightly decreases when the anodizing time increases up to 90 min. If we close up the results for argon, also the same trend for the three surface samples is observed.

3.2. Energy Accommodation Coefficient

The EAC was calculated from Equation (2) by fitting the measured heat flux with the curve of Equations (3) and (5). The EAC was obtained for each condition, i.e., four for each gas sample surface pair. The standard error of the EAC was calculated to evaluate the repeatability of the measurements. The size of the standard error could be an idea for the uncertainty of the measurement. The averaged EACs with the standard error are tabulated in Table 1. From Table 1, the EAC from Equation (3) is slightly larger than that from Equation (5), as mentioned above. Meanwhile, they are qualitatively in good agreement. Thus, only the EAC from Equation (5) is plotted with the error bars representing the standard error in Figure 4 for helium and argon.

Table 1. A table of the measured energy accommodation coefficient (EAC) with the standard error on 0 min, 30 min and 90 min sample surfaces for helium and argon.

Used Equation	Gas Species	0 Min	30 Min	90 Min
Equation (3)	Helium	0.3049 ± 0.0020	0.3582 ± 0.0051	0.3369 ± 0.0018
	Argon	0.9347 ± 0.0115	0.9675 ± 0.0039	0.9440 ± 0.0077
Equation (5)	Helium	0.3030 ± 0.0020	0.3556 ± 0.0051	0.3346 ± 0.0018
	Argon	0.9135 ± 0.0108	0.9449 ± 0.0039	0.9225 ± 0.0074

From Figure 4, the EAC appears to increase at the anodizing time of 30 min, while it decreases for 90 min. Compared with Figure 3, it is easy to understand the same trend with the gradient of the heat flux curve, since the heat flux is proportional to the EAC in the free molecular regime as in Equation (2), though it is much easier to see the difference in Figure 4. The error bars for argon seem to be much larger than those for helium. However, the absolute value of the EAC for argon is almost three times of that for helium and the relative errors are almost in the same range at less than 1.5%. Then, the measurement accuracy appears to be independent of gas species, and the heat flux is measured with reasonably good accuracy.

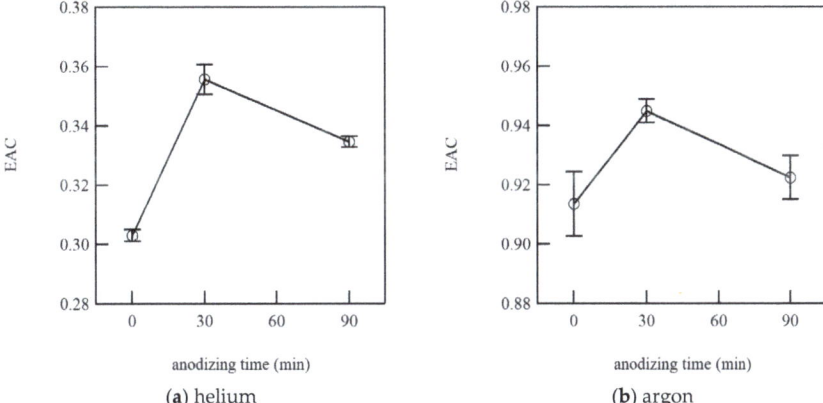

Figure 4. The measured EAC by Equation (5) as a function of the anodizing times of 0 min, 30 min and 90 min for (**a**) helium and (**b**) argon. Error bars show the standard error of 4 measurements for each condition.

Comparing the results of 0 min and 30 min sample surfaces, the EAC is clearly shown to increase by anodization of an aluminum surface. There could be two reasons for this increase: the formation of the nano-scale roughness and the oxidization of aluminum surface, i.e., the difference in materials. For 90 min sample surfaces, the surface was oxidized but is much smoother because of the dissolution of the nano-scale hexagonal cylindrical cell structure of the anodic oxide film on aluminum, as mentioned above. Therefore, we consider that the former effect accounted for the difference between the 30 min and 90 min surface samples; while, the latter effect was the difference between the 0 min and 90 min surface samples. Even though it was difficult to observe the cell structure in our SEM images in Figure 1, the nano-scale roughness was formed on the 30 min sample surfaces, and it will increase the EAC for about 0.02. This qualitative trend is coincident with the well-known characteristics of the EAC.

In [4], it is stated that the effect of macroscopic surface roughness plays only a minor role in the EAC by comparing the results on the machined and the polished surfaces for argon, nitrogen and helium. Comparing the results for the machined and polished 304 Stainless steel surfaces, the EAC decreased from 0.46 to 0.42 for helium, gave the same value of 0.87 for nitrogen, and increased from 0.95 to 0.96 for argon. In our study, the variation size in the EAC for the change of roughness was similar; however, the EAC was increased for both helium and argon. The difference between this study [4] and our study could be coming from the approach to roughening a sample surface. In [4], the machining process was employed for changing the roughness. The surface morphology was modified only for accessible areas from outside. Whereas, the anodization process, which is a wet process, was employed in our study, and it can modify the whole surface area where gas molecules approach. Therefore, the EAC increased for both gas species. It is reasonable to consider that the roughness increases the EAC, even though the size of variation is not large.

4. Conclusions

The heat flux from an anodic oxide film on aluminum was measured in the free-molecular to near free-molecular flow regime. The sample surfaces were prepared for three conditions: without anodization (0 min), and with the anodizing times of 30 min and 90 min. For our anodization conditions, 90 min was too long, and a part of the film would be dissolved. SEM images were taken for the three sample surfaces: 0 min, 30 min and 90 min. The surface seemed roughened by the anodization, but it was relatively smoothened and became powdery for 90 min sample surfaces. We failed to capture the detailed hexagonal cylindrical cell structure; however, the formation of the film was validated by an electrical resistance.

The heat flux from the sample surface was measured, and the obtained heat flux was fitted by the curve to extract the energy accommodation coefficient. The obtained energy accommodation coefficient was larger for the 30 min sample surface than for the 0 min or 90 min sample surface, indicating that the EAC increased with an increase in the surface roughness by the anodization process. The decrease of the EAC from the 30 min to 90 min sample surfaces indicated the dissolution of the anodic oxide film on aluminum, and coincided with the relatively smooth surface observed in the SEM image. These characteristics were observed for both helium and argon.

Author Contributions: Conceptualization, H.Y.; methodology, H.Y.; formal analysis, H.Y. and K.K.; investigation, H.Y. and K.K.; writing—original draft preparation, H.Y.; writing—review and editing, H.Y.; funding acquisition, H.Y. All authors have read and agreed to the published version of the manuscript.

Funding: This work was partially funded by JSPS KAKAENHI, Grant No. 18K03946.

Acknowledgments: This work was partially supported by Nanotechnology Platform Program (Nanofabrication Platform Consortium) of the MEXT, Japan. The SEM images were acquired in Nagoya University.

Conflicts of Interest: The authors declare no conflict of interest.

References

1. Goodman, F.O.; Wachman, H.Y. *Dynamic of Gas-Surface Scattering*; Academic: New York, NY, USA, 1976.
2. Saxena, S.C.; Joshi, R.K. *Thermal Accommodation and Adsorption Coefficients of Gases*; Hemisphere: New York, NY, USA, 1989.
3. Sharipov, F. *Rarefied Gas Dynamics: Fundamentals for Research and Practice*; Wiley-VCH: Berlin, Germany, 2006.
4. Trott, W.M.; Castañeda, J.N.; Torczynski, J.R.; Gallis, M.A.; Rader, D.J. An experimental assembly for precise measurement of thermal accommodation coefficients. *Rev. Sci. Inst.* **2011**, *82*, 035120. [CrossRef] [PubMed]
5. Ono, S. Structure and Growth Mechanism of Porous Anodic Oxide Films -Anodizing of Aluminum and Magnesium-. *Hyomen Kagaku* **1998**, *19*, 790–798. (In Japanese) [CrossRef]
6. Uchiyama, T.; Isoyama, E.; Otsuka, T. Surface treatment of aluminum. *J. Jpn. Inst. Light Met.* **1980**, *30*, 592–605. (In Japanese) [CrossRef]
7. Yamaguchi, H.; Imai, T.; Iwai, T.; Kondo, A.; Matsuda, Y.; Niimi, T. Measurement of thermal accommodation coefficients using a simplified system in a concentric sphere shells configuration. *J. Vac. Sci. Technol. A* **2014**, *32*, 061602. [CrossRef]
8. Yamaguchi, H.; Ho, M.T.; Matsuda, Y.; Niimi, T.; Graur, I. Conductive heat transfer in a gas confined between two concentric spheres: From free-molecular to continuum flow regime. *Int. J. Heat Mass Transf.* **2017**, *108*, 1527–1534. [CrossRef]
9. Yamaguchi, H.; Hosoi, J.; Matsuda, Y.; Niimi, T. Measurement of conductive heat transfer through rarefied binary gas mixtures. *Vacuum* **2019**, *160*, 164–170. [CrossRef]
10. Springer, G.S. Heat transfer in rarefied gases. In *Advanced in Heat Transfer*; Irvine, T.F., Harnett, J.P., Eds.; Academic Press: New York, NY, USA, 1971; Volume 7, pp. 163–218.
11. Sherman, F.S. A survey of experimental results and methods for the transitional regime of rarefied gas dynamics. In *Rarefied Gas Dynamics, Proceedings of the Third International Symposium on Rarefied Gas Dynamics, Paris, France, 26–30 June 1962*; Academic Press: New York, NY, USA, 1963; pp. 228–260.
12. Rumble, J.R., Jr.; Lide, D.R.; Bruno, T.J. *CRC Handbook of Chemistry and Physics*; CRC Press: Boca Raton, FL, USA, 2017.

© 2020 by the authors. Licensee MDPI, Basel, Switzerland. This article is an open access article distributed under the terms and conditions of the Creative Commons Attribution (CC BY) license (http://creativecommons.org/licenses/by/4.0/).

Article

Real-Time Detection of Slug Velocity in Microchannels

Salvina Gagliano [†], Giovanna Stella [†] and Maide Bucolo *,[†]

Department of Electrical, Electronics and Computer Engineering, University of Catania, v.le A. Doria 6, 95129 Catania, Italy; salvina.gagliano@unict.it (S.G.); giovanna.stella@phd.unict.it (G.S.)
* Correspondence: maide.bucolo@unict.it; Tel.: +39-095-738-2603
† These authors contributed equally to this work.

Received: 24 January 2020; Accepted: 24 February 2020; Published: 26 February 2020

Abstract: Microfluidics processes play a central role in the design of portable devices for biological and chemical samples analysis. The bottleneck in this technological evolution is the lack of low cost detection systems and control strategies easily adaptable in different operative conditions, able to guarantee the processes reproducibility and reliability, and suitable for on-chip applications. In this work, a methodology for velocity detection of two-phase flow is presented in microchannels. The approach presented is based on a low-cost optical signals monitoring setup. The slug flow generated by the interaction of two immiscible fluids {air and water} in two microchannels was investigated. To verify the reliability of the detection systems, the flow nonlinearity was enhanced by using curved geometries and microchannel diameter greater than 100 μm. The optical signals were analyzed by using an approach in a time domain, to extract the slug velocity, and one in the frequency domain, to compute the slug frequency. It was possible to distinguish the water and air slugs velocity and frequency. A relation between these two parameters was also numerically established. The results obtained represent an important step in the design of non-invasive, low-cost portable systems for micro-flow analysis, in order to prove that the developed methodology was implemented to realize a platform, easy to be integrated in a System-on-a-Chip, for the real-time slug flow velocity detection. The platform performances were successfully validated in different operative conditions.

Keywords: experimental study; optical signals monitoring; air–water flows; slug velocity; slug frequency

---

1. Introduction

Nowadays, the hydrodynamics of two-phase slug flows in microchannels [1] play an important role in the micro-nano technology, enabling the design for lab-on-chip devices in the bio-medical field as well as in chemical processes [2,3]. In this context, an open issue is that of developing detection systems, models, and control strategies easily adaptable in different operative conditions, are low-cost, able to guarantee the processes reproducibility and reliability, and suitable for on-chip applications [4,5].

The results presented in literature are strictly related to specific experimental conditions, so far from a well-established framework that can drive to the flow control. Recently, some case studies using System-on-a-Chip (SoC) predictive control strategies have been presented in literature [5,6]. The SoC offers a high control level and modularity, but its functionalities are strongly dependent on the integrated control logic and the knowledge of the process model. In this complex scenario, in which the properties of the fluids, the input flow conditions, the channel geometry and the material surface properties can strongly affect the flow dynamics, the need to develop detection systems, models, and control strategies completely independent of any constraints related to the experimental conditions represents the bottleneck for widespread diffusion that SoC uses in microfluidics applications.

In two-phase slug flows, two immiscible fluids, one dispersed in the other, are circulating in the same microsystem (for instance gas–liquid, immiscible liquid–liquid, or liquid and micro-particles) [1,7]. In general, their behavior is very complex: interfaces adopt elaborate forms and classification of regimes can sometimes lead to inextricable phase diagrams, where many regimes are mixed up, identified as bubble, slug, or plug, annular, churn, and wispy annular. Several approaches are available for a detection and control in microchannels [8], but, among them all, the optical methods have offered the advantages of a wide range of non-invasive measurement options [9]. The most common optical investigation of microfluidic flows happens by continuous monitoring using a fast Charge-Coupled detector (fast-CCD camera), or a Particle Image Velocimetry (PIV) system [10,11]. Both allow for obtaining detailed and precise flow information with the drawbacks of costly and bulky equipment. The challenge nowadays is to have methodologies based on low cost technologies easily embedded in a portable device for real-time applications. In this context, data-driven approaches based on monitoring optical signals [9] can represent a good alternative since they are non-invasive, offer an easy integration of optical sensors with the microfluidic chips [12,13], and, in future development, the possibility of being even embedded in a chip [14]. In the micro-optofluidic chip presented in [14], the advantage of the integration of micro-optical and microfluics components in one device is proved by taking the advantage of advanced signal analysis methodology to process the optical information and control the flow. Advantages can be also be envisaged for SoC applications due to the simplicity of managing optical signals, as it is proved by a wide literature on flow classification in microchannels [15,16].

In our recent works, advanced signals analysis methodologies have been developed to characterize the flow nonlinearity [17], for data-driven process identification [18,19] and to define parameters for slug flow classification [20] to be used in the development of online control strategy [21]. Starting from the results presented in [20] where the frequency analysis of signals, acquired by a low-cost monitoring setup, was used to classify the air and water dominance inside the micro-channel, in this work, the attention was focused in the development of a methodology suitable for detection of the slug velocity in micro-channels. Additionally, the developed methodology was, in a second step, implemented to realize a platform, easy to be integrated in a SoC, for the real-time slug flow velocity detection.

In the experiments presented in this paper, the slug flow was generated by the interaction of two immiscible fluids {air and water} in two serpentine microchannels of diameter {320 µm and 640 µm}. The serpentine geometry enhancing the unpredictability of the flow guarantees the robustness of the methodologies and the platform presented. The optical-based approach, previously used by the authors, based on the acquisition and processing of optical signals, has been extended to a more general framework. Two methods have been considered: the dual-slit methodology based on the cross-correlation analysis to compute the slug velocity [22] and the spectral analysis to compute the slug frequency [20]. An experimental campaign was carried out by setting nine different flow operative conditions at the inlets. Thanks to the methods used, in each experiment, it was possible to distinguish the water and air slugs velocity and the water and air slugs frequency. The two parameters were also correlated.

The results obtained were used to implement a platform for the real-time detection of the slug velocity. Two experiments were carried out monitoring the slug flow in different operative conditions. The platform performance was successfully validated. The experimental set-up and the analysis approach proposed are presented in Section 1. Then, in Section 2, the results related to data analysis of the experiments carried out are discussed in detail. Finally, in Section 3, the Platform implemented in LabVIEW is described for the real-time detection of the slug velocity showing its performance in two different experiments.

2. Materials and Methods

A continuous slug flow was generated by pumping de-ionized water and air at the Y-junction of serpentine microchannels in a Cyclic Olefin Copolymer (COC) with a square section and positioned horizontally. Two neMESYS syringe pumps were connected to the two channel inlets. The flow chart and a picture of the complete experimental set-up are shown in Figure 1a,b. The process was monitored in a microchannel area by the simultaneous acquisition of the light intensity variation using a couple of photo-diodes (sample rate of 2 kHz) placed at a distance of 10 mm from each other and a CCD (frame rate of 25 frame/s). The CCD-video was used to have a visual inspection of the process. A frame sequence of the slug passage acquired by the CCD is in Figure 1d. A detailed description of the optical set-up and the signals pre-processing is given in [20]. The serpentine microchannels used in this work are shown in Figure 2a,b, where the two rectangles are placed in the investigation areas. The microchannel in Figure 2a has a diameter of 320 µm and a length of 50 mm (labelled as G-320), and the microchannel in Figure 2b has a diameter of 640 µm and a length of 121 mm (labelled as G-640). The geometry G-320 was used in the experiments carried out to establish an optical signals processing methodology for the computation of slug velocity and slug frequency. Both geometries were considered for the validation of real-time velocity detection by the ad hoc platform implemented. In Figure 2c, an example of the trends of the signals acquired by the couple of photo-diodes $\{ph1, ph2\}$ in G-320 is shown. As discussed widely in [20], these optical signals are correlated with air/water slugs passage as follows: the top-level represents the water presence in the channel, while the lower level shows the air slugs passage. The two lowest peaks reveal the slugs fronts and rears. Indeed, during the slugs passage, the intensity of the light decreases suddenly due to the difference between the refraction index of COC ($N_{COC} = 1.5$) and air ($N_{air} = 1$), so the air slug contour becomes darker than the inside of the slug and the chip wall. This effect is less evident during the water passage since the water refraction index (Nwater = 1.3) is closer to the one of the COC. Thanks to this phenomenon, it is possible to clearly distinguish the air and water slugs passage in the signals.

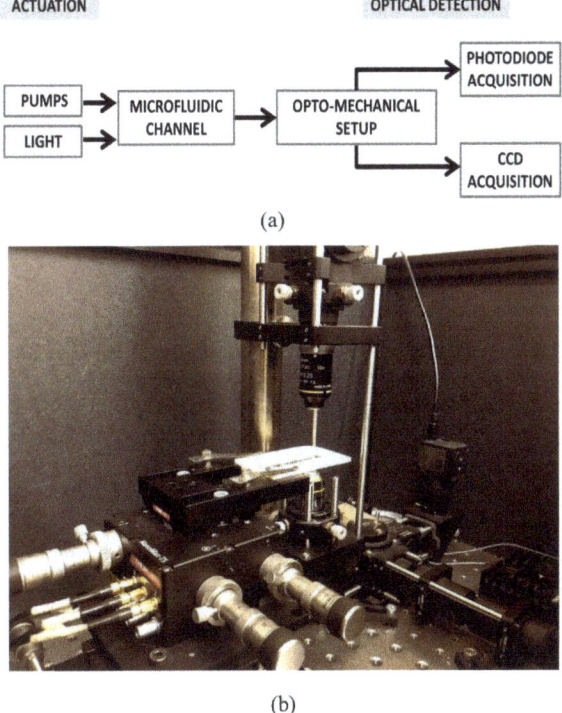

Figure 1. (a) The flowchart of the experimental set-up; (b) a picture of the opto-mechanical experimental set-up.

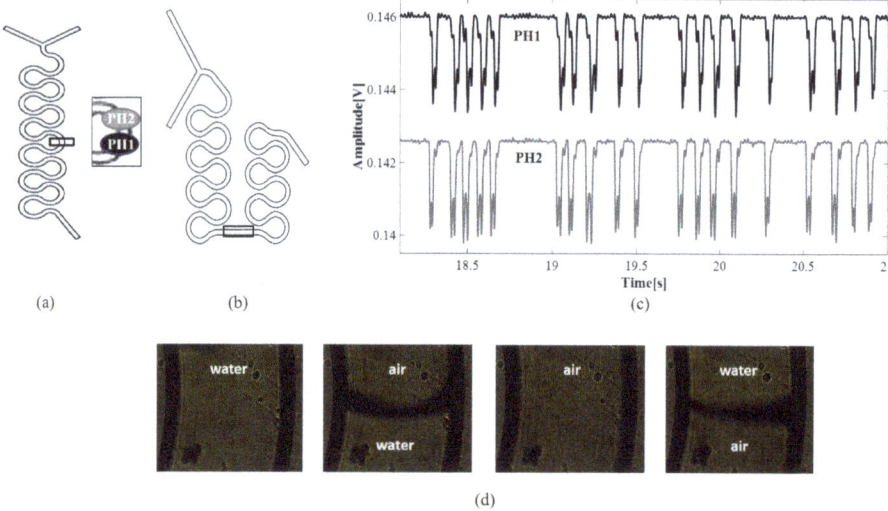

Figure 2. (**a**) the geometry of the serpentine channel of 320 μm (labelled as G-320); (**b**) the geometry of serpentine channel of 640 μm (labelled as G-640); (**c**) an example of the trends of the signals acquired by the two photo-diodes {*ph1, ph2*} in G-320 in the experiment with an input flow rate $F = 0.3$ mL/min. The two rectangles in G-320 and G-640 point out the investigation areas where the two-photo-diodes are placed; (**d**) a frame sequence related to the slug passage acquired by the CCD.

Nine experiments were carried out by feeding equal flow rates of water (F_w) and air (F_a) to the two inlets of the microchannel G-320 as in the set $F \in \{0.1, 0.2, 0.3, 0.4, 0.5, 0.6, 0.7, 0.8, 0.9\}$ mL/min. The Air Fraction (AF) was respectively $AF = 0.5$. The process was monitored in a position after three bends from the Y-junction; see the insert in Figure 2a. Data were acquired for 20 s, but the first and last 5 s of acquisition were excluded from the analysis.

Three well-known dimensionless parameters widely used in fluid dynamics to establish some basic flow characteristics were computed: the Reynolds number, the Capillary number, and the Dean number [7]. The Reynolds number information is about the transition between laminar and turbulent flow. The boundary value is debatable, but, generally in microfluidics, it is assumed to be $Re = 1$. To enhance the process nonlinearity and be able to evaluate the robustness of the methodology proposed in our experiments, its value was in the range [1, 13]. The Capillary number on the order 10^{-3}, as it is in the experiments presented, assures the slugs' formation. Finally, the curve in the serpentine geometry can determine a recirculation in the flow in the cross-section. The presence of these phenomena can be evaluated by a Dean number greater than 1, in the experiments ranges in [0.5, 6].

These parameters give indication about some characteristics of the flow taken into account: the fluids properties, the microchannel geometry, and input flow rates. Being the process highly nonlinear, it is not sufficient for its dynamical characterization and can not be used for the process control. This underlines the need of data-driven approaches that can monitor and identify the process state in real time.

To establish the velocity of the slug flow in the microchannel, the data were analyzed in time domain by using the dual-slit methodology, presented in [22]. Then, the slug frequency, corresponding to slug passage duration in time, was instigated by spectral analysis using the approach presented by the authors in [20].

2.1. The Slug Velocity by Dual-Slit Methodology

In order to evaluate the slug velocity, the cross-correlation between optical signals acquired by $\{ph1, ph2\}$ was computed for each experiment. A peak detected in the cross-correlation is representative of the time-delay of the two signals and can be correlated with the time needed by the slug to move from one investigation point $\{ph2\}$ to another one $\{ph1\}$. Knowing the distance between photo-diodes, it is possible to compute the slug velocity using the following formula:

$$v = \frac{d_s}{n * T_s} \quad (1)$$

where d_s is the distance between photo-diodes scaled based on the magnification ($d_s = 1$ mm), n is the delay in samples between the two signals, T_s is the sampling period (0.5 ms), and so $n * T_s$ is the detectable delay in time [s]. In Figure 3a, the cross-correlation function in a time window of 0.1 s is reported (equivalent to 200 *samples*) for the experiment with $F = 0.3$ mL/min. Two peaks can be detected, one at 2 ms (4 *samples*), and another one at 20 ms (40 *samples*). Based on the signal trend shown in Figure 2c, the first peak can be related to the faster sequence of air slugs that anticipates and follows the passage of a long water slug. The second peak can be related to the passage of long water slugs itself. Detecting different velocities is expectable given the process nonlinearity. Based on the values of these delays, the two velocity values obtained are {0.5 m/s, 0.05 m/s} that refer to the velocity of water and air slugs, respectively.

To validate the results, a sequence of frames obtained related to a slug passage was analyzed by the Digital Particle Image Velocimetry (DPIV) approach, used previously to detect the red blood cell velocity in microchannels in [23]. The mean velocity detected at the slug front and rear was respectively {0.034 m/s, 0.015 m/s}.

2.2. The Slug Frequency by Spectral Analysis

The spectral analysis of optical signals for the flow characterization and classification in microchannels was presented in [20] and used in [21] for the real-time slug flow control in a feed-forward configuration. The slug frequency is associated with the duration of slugs passage.

It was implemented by computing the spectrum of the optical signal $\{ph2\}$. Then, the spectrum profile was approximated with a multi-mode Gaussian model and the dominant frequencies detected. In both previous works [20,21], the attention was focused on the single-mode Gaussian model and in the highest peak. In this work, two highest peaks were considered in order to investigate the behavior of the water and air slugs. As for the cross-correlation function, the two dominant frequencies were associated with the presence of long water slugs and short air slugs. In Figure 3b, the spectrum of the optical signal $\{ph2\}$ for the experiment $F = 0.3$ mL/min is shown. The two peaks detected are at frequency $f = 1.33$ Hz and $f = 10.67$ Hz. The highest one can be associated with the passage of long water slugs $T = 0.75$ s, while the other one with the passage of short air slugs $T = 0.093$ s that refer to the frequency of water and air slugs, respectively.

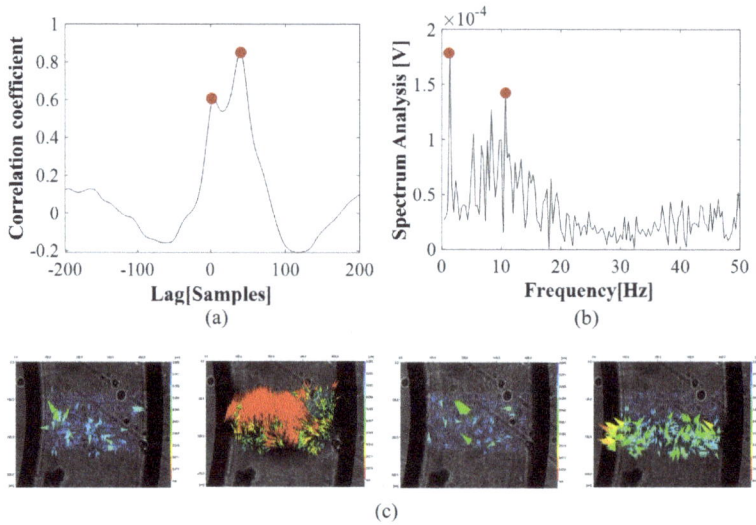

Figure 3. Optical signals processing of $\{ph1, ph2\}$ in G-320 for the experiment with an input flow rate $F = 0.3$ mL/min. (**a**) an example of the cross-correlation function for the computation of the slug velocity; (**b**) an example of the spectrum of the signal $\{ph2\}$ for the computation of the slug frequency. In both plots, the two higher peaks are highlighted with red dots; (**c**) a sequence of frames related to a slug passage analyzed by Digital Particle Image Velocimetry (DPIV).

3. Velocity and Frequency of Water and Air Slugs

Both methods were used to analyze the optical signals acquired in all the nine experiments $F \in \{0.1, 0.2, 0.3, 0.4, 0.5, 0.6, 0.7, 0.8\}$ mL/min. The signals of the two photo-diodes $\{ph1, ph2\}$ were used to compute the slug velocity by the dual-slit methodology, whereas in the case of the spectral analysis only the information of one photo-diode $\{ph2\}$ was used. Finally, the results obtained by the two approaches were mathematically correlated.

To evidence the signal dynamics in different experimental conditions, in Figure 4a, the trends of the raw optical signals for the experiments $F \in \{0.2, 0.4, 0.6, 0.9\}$ mL/min are shown. A time window of 2 s was used for $F \in \{0.2, 0.4, 0.6\}$ mL/min and 0.5 s for the experiment $F = 0.9$ mL/min. The process nonlinearity does not allow for having a periodic slugs passage; nevertheless, a changing pattern of the slug flows can be clearly distinguished at the increase of the input flow rate, as follows:

- $\{F = 0.1\text{–}0.2 \text{ mL/min}\}$ long water slugs are interlaced with short air slugs, almost one after another;
- $\{F = 0.3\text{–}0.8 \text{ mL/min}\}$ long water slugs are followed by a train of short air slugs in sequence;
- $\{F = 0.9 \text{ mL/min}\}$ a fast train of short air/water slugs.

In the experiments $F \in [0.1, \ldots, 0.8]$ characterized by the passage of long water slugs, a slower dynamic is observed than the one obtained for $F = 0.9$ where an oscillatory-like trend is recognizable.

In Figure 4b, the cross-correlation functions obtained for the four experiments $F \in \{0.2, 0.4, 0.6, 0.9\}$ mL/min are shown. A time window of 0.1 s (200 *samples*) was used for $F = 0.2$ mL/min and of 0.05 s (100 *samples*) for the other three experiments $F \in \{0.4, 0.6, 0.9\}$ mL/min. At the increase of the input flow rate, the convergence of the two peaks in one and a reduction of the time delay can be noticed, with air and water slugs having the same velocity. For the experiment $F = 0.9$ mL/min, an oscillatory trend in the cross-correlation function is evident and a sharp peak at 10 *samples* (5 ms) that stands for a velocity of 0.2 m/s.

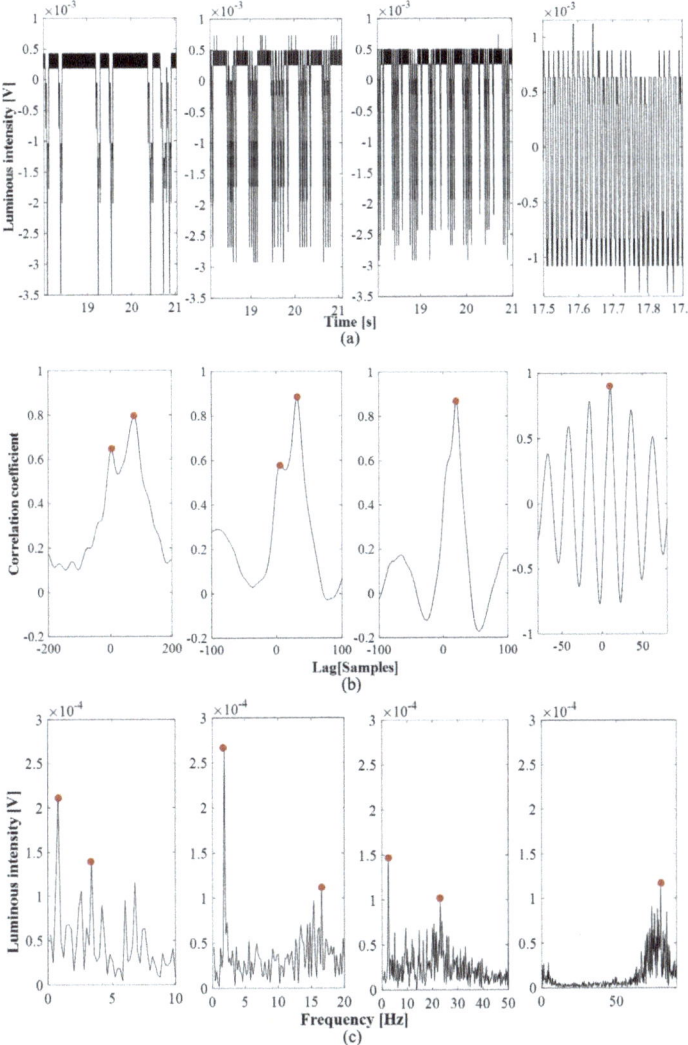

Figure 4. Dynamics of the slug flows in the experiments $F \in \{0.2, 0.4, 0.6, 0.9\}$ mL/min. (**a**) the trends of the raw optical signals acquired varying the input flow rate. The change of the flow patterns from long water slugs is evident interlaced by short air slugs one after another ($F = 0.2$ mL/min), to longer water slugs ($F = 0.4$–0.6 mL/min) followed by a train of smaller air/water passages and finally a train of smaller water/air slugs ($F = 0.9$ mL/min); (**b**) the cross-correlation between the signals acquired through the two photo-diodes, and in red dots the peaks detected; (**c**) the spectra of the optical signal $\{ph2\}$, and in red dot the peaks detected.

The cross-correlation function was computed for all the nine experiments, but two peaks were detected only in the first five $F \in \{0.1, 0.2, 0.3, 0.4, 0.5\}$ mL/min. The method does not seem to be able to differentiate the water and air velocities in the experiments $F \in \{0.6, 0.7, 0.8\}$ mL/min, even though we can notice a difference in the flow patterns of Figure 3a. Consequently, it was not possible to have an evaluation of the air velocity only in these conditions. The convergence of the slug flow towards a uniform velocity distribution, having the same water and air slugs velocity, is obtained for $F = 0.9$ mL/min.

In Figure 5a, the values of water velocity obtained by the highest peak were plotted versus the input flow rate: the blue dots are for $F \in \{0.1, 0.2, 0.3, 0.4, 0.5\}$ mL/min and the red dots for $F \in \{0.6, 0.7, 0.8, 0.9\}$ mL/min. In Figure 5b for $F \in \{0.1, 0.2, 0.3, 0.4, 0.5\}$ mL/min, the values of air velocity, obtained by the second peak, were reported. Both graphs were mathematically interpolated. The parabolic increases of the water slug velocity and the linear decreasing of the air slug velocity are worth noticing.

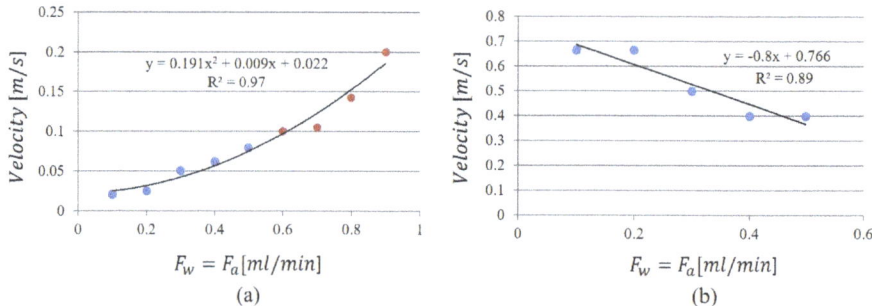

Figure 5. The cross-correlation function was computed for all the nine experiments. (**a**) the values of water velocity obtained by the highest peak were plotted versus the input flow rate: the blue dot are for $F \in \{0.1, 0.2, 0.3, 0.4, 0.5\}$ mL/min and the red dots for $F \in \{0.6, 0.7, 0.8, 0.9\}$ mL/min; (**b**) the values of air velocity, obtained by the second peak for $F \in \{0.1, 0.2, 0.3, 0.4, 0.5\}$ mL/min.

By the analysis of the slug pattern (Figure 4a), the velocity and the frequency of the air and water slugs are expected to be different from each other because of the slower dynamics and the convergence to the same value for $\{F = 0.9$ mL/min$\}$. From this perspective, the analysis in the spectral domain seems more robust.

In Figure 4c, the spectra of the optical signals for the experiments $F \in \{0.2, 0.4, 0.6, 0.9\}$ mL/min are shown. As expected, at the increase of the input flow rate, the convergence of the two peaks in one and an increase of the frequency of the slugs passage can be noticed, with the air and water slugs assuming the same duration. Coherently with the flow pattern of Figure 3a, one sharp peak is obtained in the experiment $F = 0.9$ mL/min at 79.2 Hz that stands for an average duration of the slug passage of 0.012 s.

The values of slug frequency were computed for all the nine experiments and in Figure 6 the values of the two peaks identified in the spectra are plotted versus the input flow rates $F \in \{0.1, 0.2, 0.3, 0.4, 0.5, 0.6, 0.7, 0.8\}$ mL/min. The water frequency is reported in Figure 6a, and the air frequency is reported in Figure 6b. In this case, by the mathematical interpolation at the increase of the input flow rate, the water slug frequency increases linearly in the range [0.5, 3.5] Hz, whereas the air slug frequency has a parabolic trend [1, 48] Hz. Therefore, a water slug passage can last from [0.3, 2] s and an air slug passage [0.02, 1] s. The convergence of the two dynamics for $F = 0.9$ mL/min is obtained when the high frequency becomes dominant having the water–air train flow pattern.

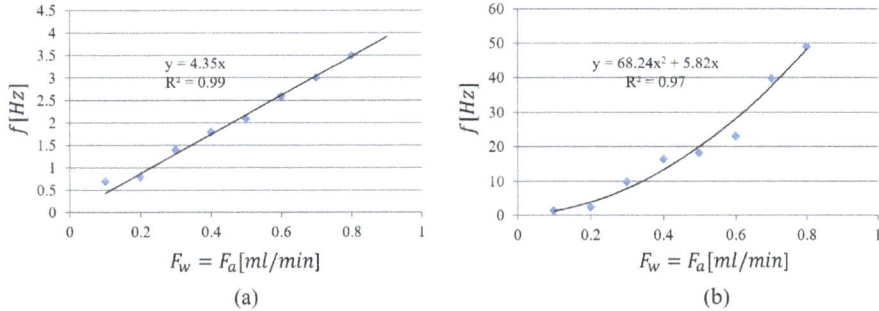

Figure 6. The values of slug frequency computed by the two peaks identified in the spectra versus the input flow rates for the eight experiments $F \in \{0.1, 0.2, 0.3, 0.4, 0.5, 0.6, 0.7, 0.8\}$ mL/min. (**a**) water frequency; (**b**) air frequency.

Finally, the results of the analysis in the frequency and time domain were graphically correlated. The spectral analysis shows two peaks for all the eight experiments $F \in \{0.1, 0.2, 0.3, 0.4, 0.5, 0.6, 0.7, 0.8\}$ mL/min before reaching the convergence for $F = 0.9$ mL/min. The cross-correlation function has two peaks for five experiments $F \in \{0.1, 0.2, 0.3, 0.4, 0.5\}$ mL/min.

In Figure 7, the water velocity was plotted versus the water frequency for $F \in \{0.1, 0.2, 0.3, 0.4, 0.5\}$ mL/min (reported as blue dots) and versus the air frequency for $F \in \{0.6, 0.7, 0.8, 0.9\}$ mL/min (reported as red dots). The points were linearly interpolated, maintaining the distinction between the input flow ranges. The nonlinearity of the process and its tendency to move from a flow pattern regime to another is underlined by a difference in the two linear interpolations. Additionally, these mathematical relations address the possibility to obtain the dominant velocity of the slug flow through the analysis of one signal in the frequency domain, thus reducing the complexity of the optical set-up and the data analysis.

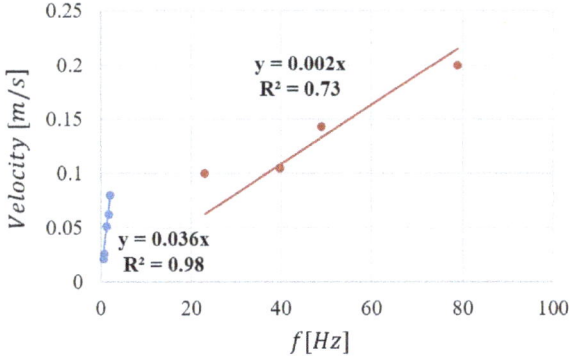

Figure 7. The water velocity plotted versus the water frequency for $F \in \{0.1, 0.2, 0.3, 0.4, 0.5\}$ mL/min (blue dots) and versus the air frequency for $F \in \{0.6, 0.7, 0.8, 0.9\}$ mL/min (red dots).

4. Platform for Real-Time Slug Velocity Detection

After the assessment of the methodology to compute the slug velocity through the analysis of optical signals, a LabVIEW Platform was implemented for a real-time slug flow monitoring. In Figure 8, a flow chart of the system is reported. A module provided by Cetoni (neMESYS SDK software) was integrated in the Platform to drive the syringe pumps. Then, the signals, acquired in cycles based on an established time window, are visualized, pre-processed and processed by using the dual-slit methodology. The velocity values obtained per cycle, for both the air and water slugs, are collected in a chart, showing the trend of the velocity versus cycle. Two experiments were carried out using the microchannels G-320 and G-640 as described in Section 2 and reported in Figure 2a,b.

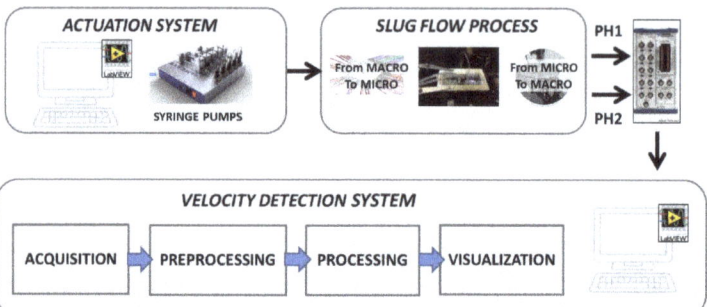

Figure 8. The flowchart of the platform to compute the slug velocity by the analysis of optical signals with the main functional blocks for the data: acquisition, pre-processing, processing, and visualization. The block that manages the pump-computer connection is also pointed out.

In Figure 9, the GUI of the Platform is presented. It is possible to distinguish two areas: the left area, used for inputting the experiment parameters, and the right area, used for the process data visualization.

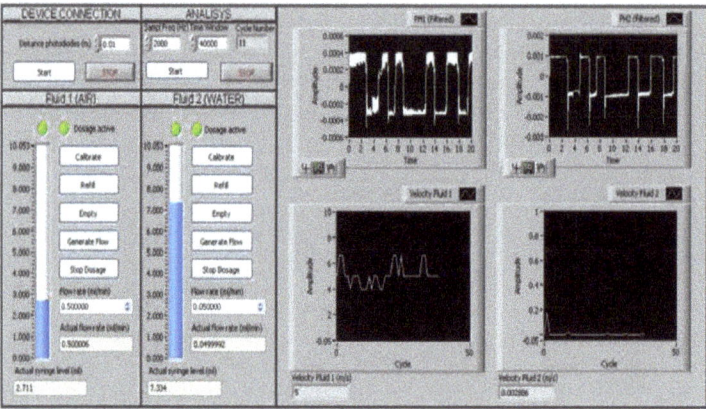

Figure 9. The GUI of the Platform. Two areas are distinguishable: the left area used for user inputting and the right area for data visualization.

- In the blank text-boxes at the top of the left area, the user can set the parameters of the experiment: the distance between the photo-diodes, the sample frequency of the acquisition board (in Hz), and the time window length to be analyzed cyclically (in *samples*). By the text-box in grey, labeled *Cycle Number*, the number of analysis cycles performed is visualized to inform the user about the number of points collected in the velocity chart and the time-horizon monitored. At the bottom of the left area, it is possible to manage the two syringe pumps, set the desired values of the input flow rates (in mL/min), and check the actual values.
- At the top of the right area, the signal acquired by the two pre-processed photo-diodes are plotted in real-time per cycle of 20 s. At the bottom, the charts that collect the water and air velocity are updated cycle-by-cycle. The values obtained for the ongoing cycle are also reported in the text-box.

The implemented algorithm includes five blocks, described in detail below.

- The first block establishes the communication computer–pumps and manages the pumps through some basic procedures as: calibration, refilling, emptying, and start-and-stop of the flow emission. In the GUI of the platform, these block functions are activated by buttons (see Figure 9). The *Start* button in the **DEVICE CONNECTION** section establishes the communication with the pumps, while, in the two **FLUID** sections, we find the buttons to *Calibrate, Refill, Empty, Generate Flow*, and *Stop Dosage* as described previously. There are also two indicators that give feedback regarding the status of the pumps: the first led indicates that the pump is connected, the second one that the dosage is active. The velocity monitoring starts by pressing the *Start* button in the **ANALYSIS** section.
- The second block carries on the two signals acquisition at the established sample frequency and time-window length that subsequently splits them up in two data vectors to be analyzed separately (*Split Signals*).
- The third block (*Filtering*), related to the pre-processing, includes the procedures for: the mean removal, a notch filter (at 50 Hz) and low pass filter. In particular, two first order Butterworth IIR filters with $\{fc = 30\ \text{Hz}\ \text{and}\ fc = 48\ \text{Hz}\}$ have been used respectively for *Ph1* and *Ph2*. The system also provides a function for saving the raw and filtered signals.
- The fourth block computes the cross-correlation between the filtered input signals (*Cross-Correlation*). Then, the *Peak Detector* function finds the delay in samples(n) related to the two peaks in the cross-correlation and computes the velocities of the two fluids by Equation 1. The system also provides a function for saving those values.
- Finally, the fifth block is for the visualization (*waveform graph*).

4.1. Slug Velocity Monitoring in a 320 µm Microchannel

The real-time flow information in the serpentine microchannel G-320 was monitored and analyzed by using a sample frequency of 2 *Khz* and in a time-window of 20 s. One inlet of the microchannel was continuously fed with a water flow rate equal to 0.05 mL/min. The air flow rate at the second inlet was varied during the experiment in the set $\{0.1, 0.3, 0.5\}$ mL/min, maintaining each value for 13 cycles, and thus was around 260 s. Differently from the experiments discussed in Section 3, the process was slowed-down by decreasing the input flow rate of the water and by setting up an unbalanced configuration of the input flow rate water–air in order to guarantee more stability in the flow velocity [20].

Figure 10a shows the trend of the air (above) and water (below) slugs velocity during the entire duration of the experiment. The red vertical dotted lines separate the three time-intervals in which the air input flow rate was maintained unaltered. The horizontal arrows report the velocity average per interval. As it can be noticed, the air and water slugs velocity was the same, while one peak was identified in the cross-correlation function. Figure 11a shows the trend of a signal $\{ph2\}$ in the condition air input flow rate 0.5 mL/min and the cross-correlation function obtained. Using very low input flow rates, no great difference is detected in the velocity values in time, but the flow velocity stability increases at the increase of the difference between the two input flow rates, as expected by previous studies [20]. The process nonlinearity leads to the variation of the velocity of 0.1 m/s that is significant considering that the total increase of the velocity is in the range $[0.3, 0.5]$ m/s. This phenomenon is particularly evident in the first two input flow rate conditions; then, a stabilization in the process is reached.

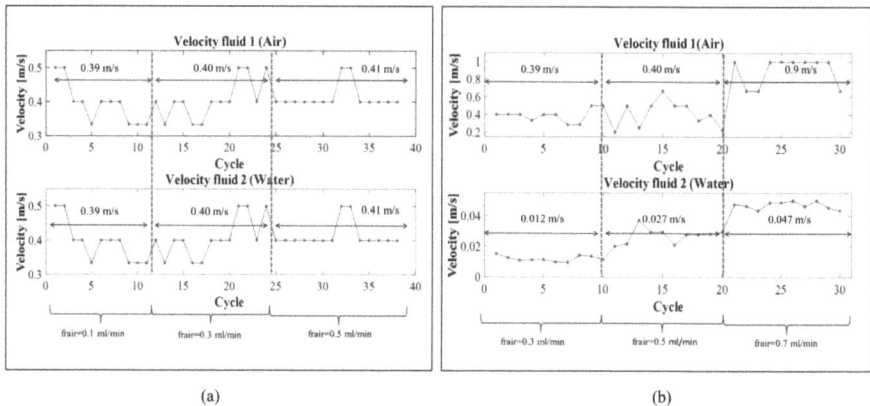

Figure 10. Velocity trends of the water and air fluids computed in real-time (**a**) in the serpentine microchannel G-320 and (**b**) in the serpentine microchannel G-640.

4.2. Slug Velocity Monitoring in a 640 µm Microchannel

The real-time flow information in the serpentine microchannel G-640 was monitored and analyzed by using a sample frequency of 2 *Khz* and in a time-window of 10 s. The input flow rate of air and water at the two inlets of the microchannel was set equal and varied during the experiment in the set $\{0.3, 0.5, 0.7\}$ mL/min, maintaining each value for 10 cycles (around 100 s). A greater diameter leads to a faster transient time, giving the possibility to reach more rapidly the flow steady condition.

Figure 10b shows the trend of the air (above) and water (below) slugs velocity during the entire duration of the experiment. The red vertical dotted lines separate the three time-intervals considered in which the two input flow rates were unchanged. The horizontal arrows report the velocity average per interval. As it can be noticed, the air and water slugs velocity are different. Figure 11b shows the trend of a signal $\{ph2\}$ in the condition input flow rate 0.5 mL/min and the two peaks in the cross-correlation function. The trends of both air and water slugs velocity increase along the entire experiment. On the other hand, the transient phase is not detectable in the first input condition $\{0.3\}$ mL/min, which is clearly visible in $\{0.5, 0.7\}$ mL/min.

In this experiment, it is important to notice that the process nonlinearity affects the air slug and water slug velocity differently. The air slug velocity increases in the range $[0.2, 1.0]$ m/s; thus, in this case, variations of 0.1 m/s have a minor impact in the velocity detection. As far as the water slug velocity is concerned, they are in the range $[0.01, 0.05]$ m/s, but, in this case, the variations detected are very small compared to the air slug velocity. The smoothness in the trend increases underlines the fact that the water flow is less affected by the process nonlinearity than the air flow.

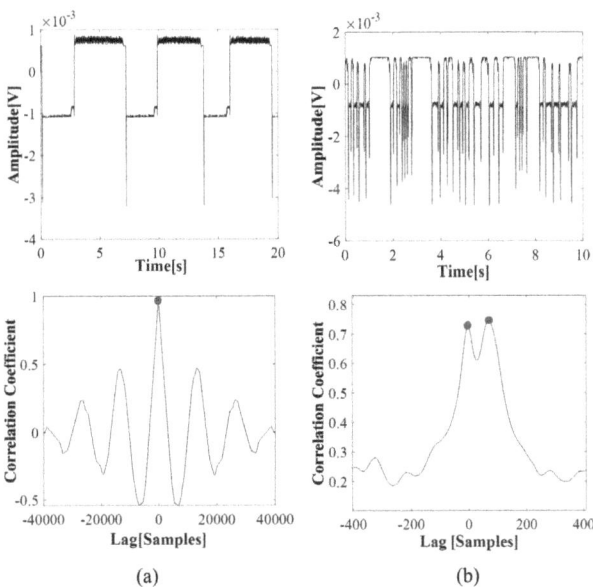

Figure 11. The trends of a signal {ph2} and the cross-correlation functions: (**a**) in the serpentine microchannel G-320 and (**b**) in the serpentine microchannel G-640.

5. Conclusions

A challenge in this work was the development of a real-time velocity detection system, for the slug flow analysis in a microchannel based on optical signals monitoring. Due to the nonlinear process, an irregular behavior is expected, so the possibility to use simple low-cost procedures for its monitoring represents an important step in the development of a microfluidic system-on-chip.

In this study, the attention was focused on two-phase flows obtained by the interaction of immiscible fluids {air and water} in two microchannels of 320 μm and 640 μm. The process was monitored through a photo-diode set-up. Two approaches based on the optical signal analysis in time and frequency domain were established and compared. The first one based on the dual-slit methodology was used to establish the water and air slugs velocity. The latter was used to detect water and air slugs frequency that is associated with the slug passage durations. A fulfilling slug flow characterization was obtained in the experimental campaigns by varying the input flow rate. In each experiment, it was possible to distinguish the water and air slugs velocity and frequency. The two parameters were also correlated.

The results obtained were used to implement a platform in LabVIEW for the real-time detection of the slug velocity. Two experiments were carried out monitoring the slug flow in different operative conditions using two serpentine microchannels of diameter {320 μm and 640 μm}. The platform performances were successfully validated.

The results obtained represent an important step in the development of non-invasive, low-cost portable systems for micro-flow analysis which could also be suitable for an easy on-chip integration. In future developments, the analysis in frequency domain will be integrated in the platform and its performances will be tested by using different two-phase slug flows.

Author Contributions: The authors contributed equally to this work in conceptualization, methodology, software and writing. All authors have read and agreed to the published version of the manuscript.

Funding: This research was funded by FONDI PER LA RICERCA DI ATENEO- PIANO PER LA RICERCA 2019.

Conflicts of Interest: The authors declare no conflict of interest.

References

1. Serizawa, A.; Feng, Z.; Kawara, K. Two-phase flows in microchannel. *Exp. Therm. Fluid Sci.* **2002**, *26*, 703–714. [CrossRef]
2. Mashaghi, S.; Abbaspourrad, A.; Weitz, D.A.; van Oijen, A.M. Droplet microfluidics: A tool for biology, chemistry and nanotechnology. *Trends Anal. Chem.* **2016**, *82*, 118–125. [CrossRef]
3. Whitesides, G.M. The origins and the future of microfluidics. *Nature* **2006**, *442*, 368–373. [CrossRef] [PubMed]
4. Bleris, L.G.; Vouzis, P.D.; Garcia, J.G.; Arnold, M.G.; Kothare, M.V. Pathways for optimization-based drug delivery. *Control Eng. Pract.* **2007**, *15*, 1280–1291. [CrossRef]
5. Bleris, L.G.; Garcia, J.; Kothare, M.V.; Arnold, M.G. Towards embedded model predictive control for System-on-a-Chip applications. *J. Process Control* **2006**, *16*, 255–264. [CrossRef]
6. Maddala, J.; Rengaswamy, R. Droplet digital signal generation in microfluidic networks using model predictive control. *J. Process Control* **2013**, *23*, 132–139. [CrossRef]
7. Tabeling, P. *Introduction to Microfluidics*; Oxford University Press: New York, NY, USA, 2005.
8. Bucolo, M.; Guo, J.; Intaglietta, M.; Coltro, W. Guest Editorial, Special Issue on Microfluidics Engineering for Point-of-Care Diagnostics. *IEEE Trans. Biomed. Circuits Syst.* **2017**, *11*, 1488–1499. [CrossRef]
9. Kuswandi, B.; Nuriman, J.; Huskens, W.; Verboom, W. Optical sensing systems for microfluidic devices: A review. *Anal. Chim. Acta* **2007**, *601*, 141–155. [CrossRef]
10. Seol, D.G.; Bhaumik, T.; Bergmann, C.; Socolofsky, S.A. Particle Image Velocimetry Measurements of the Mean Flow Characteristics in a Bubble Plume. *J. Eng. Mech.* **2007**, *133*, 665–676. [CrossRef]
11. Qu, J.W.; Murai, Y.; Yamamoto, F. Simultaneous PIV/PTV Measurements of Bubble and Particle Phases in Gas-liquid Two-phase Flow Based on Image Separation and Reconstruction. *J. Hydrodyn.* **2004**, *16*, 756–766.
12. Kraus, T.; Gunther, A.; de Mas, N.; Schmidt, M.A.; Jensen, K.F. An integrated multiphase flow sensor for microchannels. *Exp. Fluids* **2004**, *36*, 819–832. [CrossRef]
13. Sapuppo, F.; Llobera, A.; Schembri, F.; Intaglietta, M.; Cadarso, V.J.; Bucolo, M. A polymeric micro-optical interface for flow monitoring in biomicrofluidics. *Biomicrofluidics* **2010**, *4*, 1–13. [CrossRef] [PubMed]
14. Cairone, F.; Gagliano, S.; Carbone, D.C.; Recca, G.; Bucolo, M. Micro-optofluidic switch realized by 3D printing technology. *Microfluid. Nanofluid.* **2016**, *20*, 61–71. [CrossRef]
15. Mahvash, A.; Ross, A. Application of CHMMs to two-phase flow patterns identification. *Eng. Appl. Artif. Intell.* **2008**, *21*, 1142–1152. [CrossRef]
16. Mahvash, A.; Ross, A. Two-phase flow pattern identification using continuous hidden Markov model. *Int. J. Multiph. Flow* **2008**, *34*, 303–311. [CrossRef]
17. Schembri, F.F.; Sapuppo, F.; Bucolo, M. Experimental classification of nonlinear dynamics in microfluidic bubbles flow. *Nonlinear Dyn.* **2012**, *67*, 2807–2819. [CrossRef]
18. Cairone, F.; Bucolo, M. Data-driven identification of two-phase microfluidic flows. In Proceedings of 24th Mediterranean Conference on Control and Automation (MED), Athens, Greece, 21–24 June 2016.
19. Cairone, F.; Anandan, P.; Bucolo, M. Nonlinear systems synchronization for modeling two-phase microfluidics flows. *Nonlinear Dyn.* **2018**, *92*, 75–84. [CrossRef]
20. Cairone, F.; Gagliano, S.; Bucolo, M. Experimental study on the slug flow in a serpentine microchannel. *Int. J. Exp. Therm. Fluid Sci.* **2016**, *76*, 34–44. [CrossRef]
21. Gagliano, S.; Cairone, F.; Amenta, A.; Bucolo, M. A Real Time Feed Forward Control of Slug Flow in Microchannels. *Energies* **2019**, *12*, 2556. [CrossRef]
22. Sapuppo, F.; Bucolo, M.; Intaglietta, M.; Johnson, P.C.; Fortuna, L.; Arena, A. An improved instrument for real-time measurement of blood velocity in microvessels. *IEEE Trans. Instrum. Meas.* **2007**, *56*, 2663–2671. [CrossRef]
23. Cairone, F.; Ortiz, D.; Cabrales, P.J.; Intaglietta, M.; Bucolo, M. Emergent behaviors in RBCs flows in micro-channels using digital particle image velocimetry. *Microvasc. Res.* **2018**, *116*, 77–86. [CrossRef] [PubMed]

© 2020 by the authors. Licensee MDPI, Basel, Switzerland. This article is an open access article distributed under the terms and conditions of the Creative Commons Attribution (CC BY) license (http://creativecommons.org/licenses/by/4.0/).

Article

A Hybrid Numerical Methodology Based on CFD and Porous Medium for Thermal Performance Evaluation of Gas to Gas Micro Heat Exchanger

Danish Rehman [1],*, Jojomon Joseph [2,3], Gian Luca Morini [1], Michel Delanaye [3] and Juergen Brandner [2]

1. Microfluidics Laboratory, Department of Industrial Engineering (DIN), University of Bologna, 40131 Bologna, Italy; gianluca.morini3@unibo.it
2. Institute of Microstructure Technology (IMT), Karlsruhe Institute of Technology, D-76344 Eggenstein-Leopoldshafen, Germany; joseph.jojomon@mitis.be (J.J.); juergen.brandner@kit.edu (J.B.)
3. MITIS SA, Rue del Rodje Cinse 98, 4102 Seraing, Belgium; michel.delanaye@mitis.be
* Correspondence: danish.rehman2@unibo.it

Received: 24 January 2020; Accepted: 17 February 2020; Published: 20 February 2020

Abstract: In micro heat exchangers, due to the presence of distributing and collecting manifolds as well as hundreds of parallel microchannels, a complete conjugate heat transfer analysis requires a large amount of computational power. Therefore in this study, a novel methodology is developed to model the microchannels as a porous medium where a compressible gas is used as a working fluid. With the help of such a reduced model, a detailed flow analysis through individual microchannels can be avoided by studying the device as a whole at a considerably less computational cost. A micro heat exchanger with 133 parallel microchannels (average hydraulic diameter of 200 µm) in both cocurrent and counterflow configurations is investigated in the current study. Hot and cold streams are separated by a stainless-steel partition foil having a thickness of 100 µm. Microchannels have a rectangular cross section of 200 µm ×200 µm with a wall thickness of 100 µm in between. As a first step, a numerical study for conjugate heat transfer analysis of microchannels only, without distributing and collecting manifolds is performed. Mass flow inside hot and cold fluid domains is increased such that inlet Reynolds number for both domains remains within the laminar regime. Inertial and viscous coefficients extracted from this study are then utilized to model pressure and temperature trends within the porous medium model. To cater for the density dependence of inertial and viscous coefficients due to the compressible nature of gas flow in microchannels, a modified formulation of Darcy–Forschheimer law is adopted. A complete model of a double layer micro heat exchanger with collecting and distributing manifolds where microchannels are modeled as the porous medium is finally developed and used to estimate the overall heat exchanger effectiveness of the investigated micro heat exchanger. A comparison of computational results using proposed hybrid methodology with previously published experimental results of the same micro heat exchanger showed that adopted methodology can predict the heat exchanger effectiveness within the experimental uncertainty for both cocurrent and counterflow configurations.

Keywords: reduced model; LMTD method; conjugate heat transfer (CHT); compressible fluid; maldistribution

1. Introduction

Micro heat exchangers (µHx) are becoming of great interest for applications where portability, high energy efficiency and ultra high heat transfer rates are required such as the case with microelectronics cooling. Typically, µHxs are composed of multiple layers (similar to plate type heat exchangers) where each layer contains a large number of parallel microchannels (MCs). Detailed CFD modeling of these devices therefore, requires a significant amount of computational power. This computational burden can be reduced by modeling the MCs as a porous medium. Originally proposed for estimating pressure drop of incompressible flow over a bed of spheres (porous medium), Darcy's Law has been extended to multiple parallel MCs in heat sinks. An analytical model was first developed by Kim et al. [1,2] where they modeled the MCs as porous media and compared modeling results with experimental results of Tuckerman & Pease [3] as well as with Knight et al. [4]. Results showed that the developed model can be used for thermal performance and optimization of MC heat sinks. The same model has been studied by Liu and Garimella [5] and improved by Lim et al. [6]. Porous media-based analytical models tend to solve a three equations model for fluid flow and heat transfer through MC heat sinks using simplified momentum and energy equations. A porous medium-based computational model was validated against experimental studies of three ceramic µHxs by Alm et al. [7]. They found out that the heat transfer estimated by the porous model is lower than the experimental results and associated it to the channel blockage effects present in the experimental devices that are nearly impossible to cater for, in the numerical model. A common trait of all the studies conducted for MC heat sinks/µHxs is the use of an incompressible fluid. A porous medium approximation of a compact heat exchanger used in a micro gas turbine application has recently been presented by Joseph et al. [8–10]. Channel dimensions and operating pressure were such that gas was incompressible whereas the operating temperature of the hot fluid was higher than 1000 K. A porous model approximation for a double layered µHx where gas experiences strong compressibility effects, has recently been presented by the authors [11], for parallel flow arrangement. In the current study, the previously reported work is extended to cover the counterflow arrangement as well and detailed comparisons between experimental and numerical results for both flow configurations are being presented.

In this work, a two step methodology is proposed to conduct a performance evaluation study on µHx with acceptable computational power. A gas to gas double layer µHx as shown in Figure 1 that has been experimentally investigated previously by Yang et al. [12,13] and Gerken et al. [14], is used for validation of proposed methodology.

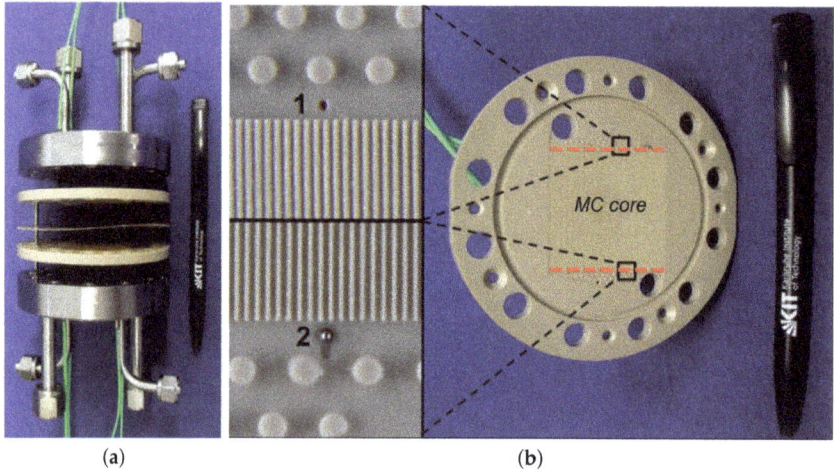

Figure 1. Experimental assembly for double layer Micro heat exchangers (µHx) (taken from [15]) (**a**), and zoomed view of single layer (**b**).

As a first step, a 3D conjugate heat transfer (CHT) model of gas to gas µHx is developed without distributing and collecting manifolds. Resulting pressure, velocity and temperature fields are utilized to calculate inertial and viscous coefficients of the modified Darcy's porous law. A complete single layer of µHx with manifolds is then modeled with boundary conditions such that MCs are modeled as a porous medium with a low resolution mesh. This is achieved by modeling the required pressure drop as a momentum source term using inertial and viscous coefficients of the porous medium and applying a free slip boundary condition at all MC walls. Similarly to incorporate heat transfer in MCs core, a source term derived by CHT analysis of MCs only, is also introduced in the complete single layer model.

2. Background

Pressure drop (Δp) of a fluid through a porous medium of length L can be expressed using an extended Darcy–Forchheimer (here after referred to as simply Darcy) law as follows:

$$\frac{-\Delta p}{L} = \frac{\mu \dot{G}}{\alpha \rho} + \frac{\Gamma \dot{G}^2}{2\rho} \tag{1}$$

where $\frac{1}{\alpha}$ is viscous coefficient representing porous medium permeability and Γ is the inertial coefficient of the Darcy's law, ρ and μ denote density and dynamic viscosity of the fluid respectively, and $\dot{G}$ denotes mass flow rate ($\dot{m}$) per unit area A ($\dot{G} = \dot{m}/A$). Calculation of viscous ($\frac{1}{\alpha}$) and inertial (Γ) coefficients is usually done using experimental pressure drop results and therefore various empirical relations exist for different porous media geometries. No such experimental relations exist for the gas flows in µHx, however. As mentioned earlier, in this work porous medium coefficients are extracted from CHT analysis using a modified Darcy law. Thus, integrating Equation (1) in the streamwise direction 'x' of the MC yields:

$$\int_{in}^{out} \frac{-\Delta p}{L} dx = \frac{\mu \dot{G}}{\alpha} \int_{in}^{out} \frac{dx}{\rho} + \frac{\Gamma \dot{G}^2}{2} \int_{in}^{out} \frac{dx}{\rho} \tag{2}$$

Differentiating the ideal gas law resuts in:

$$dp = RT d\rho + \rho R dT \tag{3}$$

Utilizing the definition of speed of sound i.e., $\frac{dp}{d\rho} = c^2$, above equation can be rewritten as:

$$d\rho \left(c^2 - RT\right) = \rho R dT \tag{4}$$

Assuming that temperature change is linear along the length of the MC, following can be obtained:

$$dT = \frac{\Delta T}{L} dx \tag{5}$$

Substituting Equation (5) into Equation (4):

$$dx = \frac{L\left(c^2 - RT\right)}{R\Delta T} \left[\frac{d\rho}{\rho}\right] \tag{6}$$

Substituting Equation (6) into Equation (2) finally results:

$$\frac{-\Delta p}{L} \int_{in}^{out} \frac{d\rho}{\rho} = \frac{\mu \dot{G}}{\alpha} \int_{in}^{out} \frac{d\rho}{\rho^2} + \frac{\Gamma \dot{G}^2}{2} \int_{in}^{out} \frac{d\rho}{\rho^2} \tag{7}$$

Integrating Equation (7) between inlet 'in' and outlet 'out' of the MC yields:

$$\frac{-\Delta p}{L} = \mu \zeta \dot{G} \left(\frac{1}{\alpha}\right) + \frac{\zeta \dot{G}^2}{2} (\Gamma) \tag{8}$$

where $\zeta = \frac{1}{\ln\left(\frac{\rho_{out}}{\rho_{in}}\right)}\left(\frac{1}{\rho_{in}} - \frac{1}{\rho_{out}}\right)$.

Boundary conditions of CHT analysis are chosen such that there are no heat losses to the surroundings, therefore all the heat loss by hot fluid must be shared between partition foil (solid wall) and cold fluid. Therefore porous medium coefficients can be extracted from either side of the CHT model. In this work, inertial and viscous coefficients are extracted from the channel with hot fluid. From a CHT analysis of MCs only, all parameters in the Equation (8) are available for specific mass flow, with inertial and viscous coefficients as the only two unknowns. However, if Equation (8) is applied to two consecutive mass flows $\dot{m}_i$ and $\dot{m}_{i+1}$, an average value of both the coefficients between these mass flows can be found out by solving a system of two linear equations for two unknowns. This is repeated for the whole range of mass flow being studied and a polynomial fit on these evaluated coefficients is used as an input for the porous model. Similarly, a volumetric heat loss of hot fluid channel is calculated as a function of mass flow from CHT analysis using:

$$\dot{q}_v = \frac{\dot{m}C_p\Delta T}{AL} \qquad (9)$$

where C_p is the specific heat of fluid at constant pressure. A polynomial fit onto this volumetric heat loss is used as a source term in energy equation while solving for the porous model. Utilizing the results of the CHT model, a MATLAB program is used to implement the above mentioned strategy for the extraction of the porous medium coefficients and the heat source term.

3. Numerical Methodology

As described earlier, two different numerical setups are used in the current work. First a 3D CHT of MCs only, without distributing and collecting manifolds is performed. Secondly, using the porous medium coefficients and the heat source term derived from CHT analysis, the performance of the complete heat exchanger with manifolds is analyzed. Reynolds number at the inlet of MC is defined by:

$$Re = \frac{\dot{m}D_h}{\mu A} \qquad (10)$$

where hydraulic diameter (D_h) of a rectangular MC with width (w) and height (h) is defined as:

$$D_h = \frac{2wh}{w+h} \qquad (11)$$

From CHT analysis, heat transfer rate ($\dot{Q}$) on hot (h) and cold (c) side can be defined by using respective flow quantities as:

$$\dot{Q} = \dot{m}C_p\Delta T \qquad (12)$$

Finally heat exchanger effectiveness, defined as the ratio of actual heat transfer rate and maximum potential heat transfer rate available, can be calculated using:

$$\varepsilon = \frac{\dot{Q}_{av}}{(\dot{m}C_p)_{min}(T_{h,in} - T_{c,in})} \qquad (13)$$

where $\dot{Q}_{av} = \frac{\dot{Q}_c + \dot{Q}_h}{2}$, is the average value of the heat transfer evaluated on the cold and hot side of µHx. For porous model of µHx, heat exchanger effectiveness to be compared with experimental results, is calculated for the MC core only. This essentially means that temperature difference from the inlet to outlet of MC core is used to calculate ε. Moreover, as only one layer is computationally modeled in porous model, resulting effectiveness is calculated as follows:

$$\varepsilon_p = \frac{\dot{Q}_p}{(\dot{m}C_p)_p(T_{h,in} - T_{c,in})} = \frac{\overline{T}_{MC,in} - \overline{T}_{MC,out}}{T_{h,in} - T_{c,in}} \qquad (14)$$

where subscript 'p' denotes the porous model. $\overline{T}_{MC,in}$ and $\overline{T}_{MC,out}$ denote the mass flow weighted averages of static temperatures, at the inlet and outlet of N number of MCs, respectively. The parameter to quantify mass flow maldistribution from the Porous model is defined as:

$$\dot{m}_{dev} = \left(\frac{\dot{m}_i - \dot{m}_{ideal}}{\dot{m}_{ideal}} \right) \times 100 \tag{15}$$

where $\dot{m}_i$ denotes the mass flow through ith MC out of total (N) MCs. Whereas, ideal mass flow rate ($\dot{m}_{ideal}$) can be calculated using Equation (10) for a given Re.

In order to extract inertial and viscous coefficients for the porous medium using methodology outlined earlier, a 3D CHT model is setup where only MCs are modeled without considering both collecting and distributing manifolds. However, to allow any possible underexpansion at the outlet of hot and cold MCs, computational domain is extended $15D_h$ in the streamwise and $32D_h$ in the lateral direction as shown in Figure 2. A meshed model for a complete layer of µHx is also shown in Figure 3. Geometry and meshing for both models are done using Design Modeler and ANSYS Meshing software respectively. A mesh of $40 \times 40 \times 100$ is used in the MCs for CHT analysis whereas a coarse mesh of $3 \times 3 \times 40$ is used for MCs in case of porous model. Mesh is refined near the walls to capture any flow vortices present in the model. A commercial solver CFX based on finite volume methods is used for the flow simulations. Ideal Nitrogen gas is used as working fluid for both models. Simulation relevant parameters used in analyses are tabulated in Table 1.

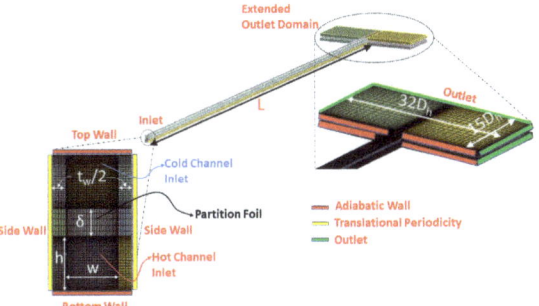

Figure 2. Mesh and geometric details for co-current conjugate heat transfer (CHT) analysis.

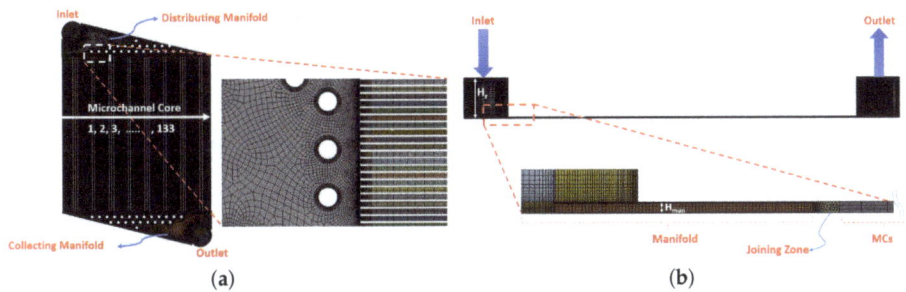

Figure 3. Geometry and mesh details of porous model: top view (**a**), and side view (**b**).

Laminar flow solver is used for the CHT model whereas a transient turbulence model $\gamma - Re_\theta$ [16–18] is utilized for the porous model. Higher order advection scheme available in CFX is utilized and pseudo time marching is done using a physical timestep of 0.01 s. A convergence criteria of 10^{-6} for RMS residuals of governing equations is chosen while monitor points for pressure and velocity at the MC inlet and outlet are also observed during successive iterations. If residuals stayed higher than supplied criteria, the solution is deemed converged when monitor points did not show any variation for 200 consecutive iterations. Reference pressure of 101 kPa was used for the simulation and all the other pressures are defined with respect to this reference pressure. Energy equation was activated using Total energy option available in CFX which adopts energy equation without any simplifications in governing equations solution. Kinematic viscosity dependence on gas temperature is defined using Sutherland's law. Further details of boundary conditions used in CHT and Porous model can be found in Tables 2 and 3 respectively.

Table 1. μHx geometry used for simulations.

Parameter	Symbol (Units)	Value
MC width	w (μm)	200
MC height	h (μm)	200
MC Length	L (μm)	40
Hydraulic Diameter	D_h (μm)	200
Wall Thickness	t_w (μm)	100
MC housing (PMMA) conductivity	k_{MC} (W/m/K)	0.25
Partition Foil (Stainless Steel) thickness	δ (μm)	100
Partition Foil conductivity	k_{PF} (W/m/K)	15

Table 2. Boundary conditions used in the CHT Analysis.

Boundary	Value	
	Hot Side	Cold Side
Inlet	$-\dot{m}$ evaluated using Equation (10) for cold side	
	$-T_{h,in} = 90\ °C$	$-T_{c,in} = 20\ °C$
Side Walls	Translational Periodicity	
Top & Bottom Walls	Adiabatic	
Outlet	Pressure outlet, p = p_{atm}	

Table 3. Boundary conditions used in the porous model for μHx.

Boundary	Value
Inlet	$-\dot{m}$ from experimental testing
	$-T_{h,in} = 90\ °C$
MCs walls	Free slip
Inertial and viscous coefficients	Determined from CHT analysis
Energy source term	Determined from CHT analysis
Manifolds walls	Adiabatic/ No slip
Outlet	Pressure outlet, p = p_{atm}

Using the CHT model, global as well as local evolution of flow variables with inlet mass flow is evaluated at six different cross-sectional planes defined at x/L of 0.005, 0.1, 0.5, 0.9, 0.95 and 0.995 respectively. In addition, two planes defined at x/L of 0.0001 and 0.9995 are treated as the inlet and outlet of MC, respectively. Results from these planes for both hot and cold fluid sides are further post processed in MATLAB to deduce required flow quantities. Thermal effectiveness is then simply evaluated using Equation (13). Once the porous medium coefficients namely inertial (Γ) and viscous ($\frac{1}{\alpha}$) are determined using CHT model of a double layer μHx, the next step is to setup a complete single layer porous model with inlet and outlet manifolds.

4. Results

4.1. CHT Model

CHT model with linear periodicity at side walls represents an ideal situation where there exists no maldistribution for parallel MCs. This essentially means that all parallel MCs would have the same mass flow at their respective inlets and manifold does not play a significant role in the performance evaluation. Results for the heat transfer rate for both cocurrent and counterflow configurations are shown in Figure 4. For an incompressible fluid, the heat transfer rate using the numerical model on both sides should be equal. But for gases, as the gas flow experiences additional acceleration due to compressibility, heat gain on the cold side tends to differ from heat loss on the hot side. An interesting fact is that the heat transfer rate on the hot side keeps on increasing with increasing mass flow. On the contrary, it keeps on decreasing on the cold side. This holds for both cocurrent and counterflow configurations. Therefore even though there are no losses modeled to the surroundings in the CHT model, due to compressibility effects gas flows still exhibit a difference in heat transfer rate between hot and cold sides.

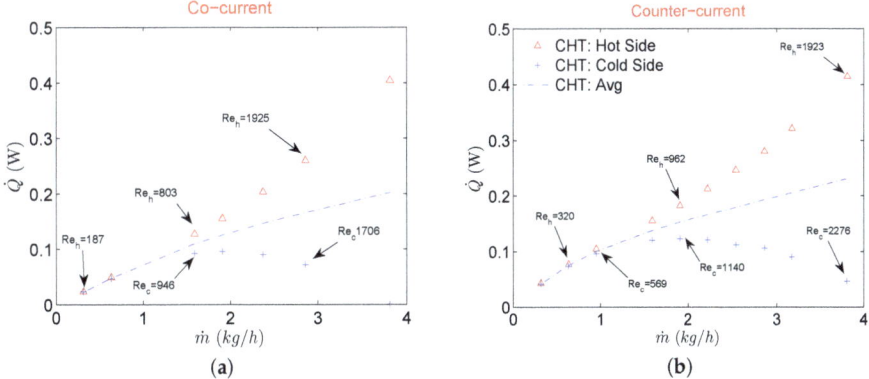

Figure 4. Heat transfer rate for CHT analysis when flow configuration is cocurrent (**a**), and counterflow (**b**).

Similar behavior can also be seen in effectiveness where it increases with the mass flow for hot fluid and decreases for the cold fluid as shown in Figure 5.

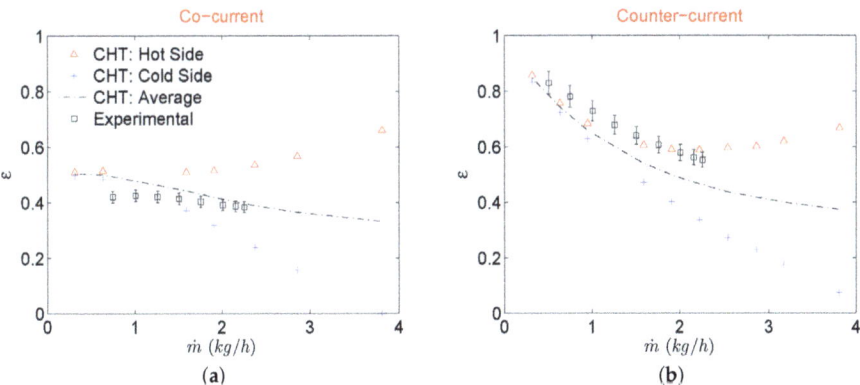

Figure 5. Heat exchanger effectiveness for CHT analysis when flow configuration is cocurrent (**a**), and counterflow (**b**).

This is simply because gas flow accelerates at the expense of kinetic energy, therefore a part of total energy is utilized for acceleration. Therefore for gas to gas μHxs, it is not recommended to operate in high mass flow regimes because at higher mass flows, gas flows experience a "self cooling" phenomenon at the expense of higher pressure drop. This phenomenon is evident in Figure 6 which shows the evolution of the temperature at the centerline of both fluid streams and partition foil (solid) along the direction of the flow. For smaller mass flows (Re) temperature profiles of hot and cold streams follow the typical trend for incompressible fluids under cocurrent configuration where the temperature of hot and cold fluids is symmetric and partition foil assumes an average constant temperature of the two sides. As fluid velocities increase at the inlets of respective streams, profile symmetry deteriorates with partition foil assuming a temperature that is more influenced by the hot side than the cold fluid side. The outlet temperature of the hot fluid keeps on decreasing with an increased mass flow rate signaling a better heat transfer process. On the contrary, if one is to look at the outlet temperature of the cold fluid a significant decrease is observed by increasing the mass flow rate. This signifies that the cold fluid stream is utilizing the transferred thermal energy from the hot stream, only to increase the velocity (kinetic energy) [19–22].

Similarly, a higher decrease of temperature and pressure very close to the outlet of the MC for the hot fluid stream is also due to a sudden expansion of the gas due to compressibility. As a result of this strong compressibility effect, gases in both fluid streams are utilizing respective thermal energies and converting them in kinetic energy thereby deteriorating the overall heat transfer process. Temperature decrease of the cold stream is such that the static temperature at the outlet is lower than the inlet value showing no active participation of cold fluid stream into the overall heat exchange of the device at higher mass flow rates. This also explains the decrease of $\dot{Q}$ in Figure 4 on the cold side of the μHx and a continuous increase of $\dot{Q}$ on the hot side with increasing mass flow rates. For data reduction of the most experimental investigations, an average value of $\dot{Q}$ is used to calculate the overall heat transfer coefficient (U) to further evaluate the heat exchanger effectiveness (ε). This is done due to the practical limitation of the heat losses to the surroundings in the laboratory environment from both streams. A similar approach has been used by Yang et al. [12,13,15] and Koyama et al. [19,20] for gas to gas μHxs. There are heat losses to the surroundings in a typical experimental campaign but as shown in this work, given a good insulation of the μHx, the real reason for the deviation of $\dot{Q}$ on hot and cold fluid streams in a gas to gas μHx might actually stems from gas compressibility. However, to compare the current numerical results with the experimental results of Yang et al. [13], ε is calculated with an average $\dot{Q}_{ave}$ from CHT analysis as it was done in experimental results. The final comparison is shown in Figure 5. For a cocurrent configuration, a CHT analysis overestimates the average ε as compared to experimental results for all the mass flows considered in this study. Difference between the two decreases though at higher mass flow rates. On the contrary, ε for counterflow configuration is underestimated by CHT analysis for the complete range of mass flows investigated. Such discrepancies between experimental results of μHx and an equivalent CHT analysis with periodic boundary conditions are expected as CHT model used in this study is devoid of flow maldistribution effects. Also, any temperature change that might occur in the manifolds due to flow deceleration caused by the presence of numerous circular pillars in distributing and collecting manifolds is not catered for.

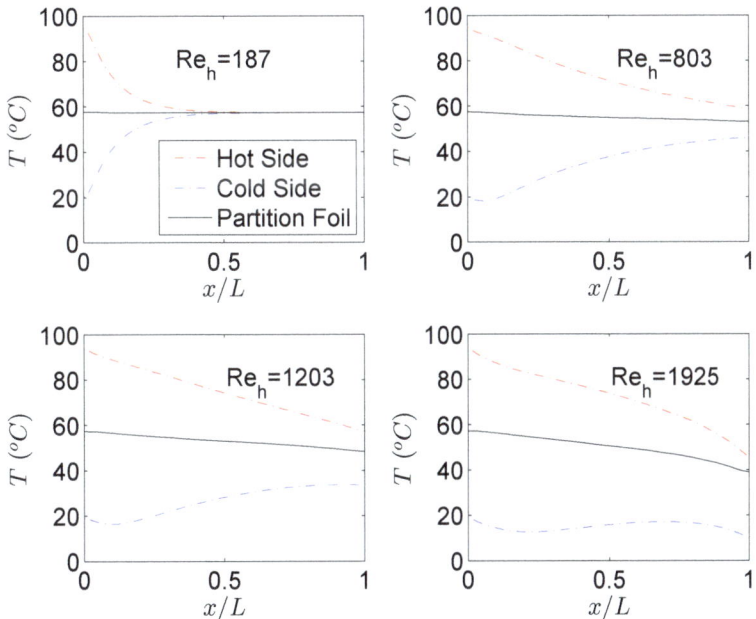

Figure 6. Temperature along the length of the hot and cold MCs for various Re in cocurrent flow configuration.

Further step is to evaluate the inertial and viscous coefficients of the modified Darcy's law using flow quantities evaluated in CHT analysis. For this reason, flow characteristics are extracted at various planes along the length of the hot and cold fluid MCs. These are used to form a set of linear systems of equations to solve for porous medium coefficients utilizing the methodology outlined in Section 1. The solutions of these systems of equations for various mass flow rates results in data points where a polynomial with mass flow rate as dependent variable can be fit to be given as input to the porous medium model available in ANSYS CFX®. In theory, for the range of inlet gas temperatures considered in this study pressure drop from hot side and cold side should be almost similar and furthermore it should be independent of the flow configuration. Therefore any side of the fluid stream (hot or cold) from CHT analysis can be used to evaluate porous medium coefficients. For the scope of this work, hot fluid side is used to evaluate porous medium coefficients. Resulting viscous and inertial coefficients are shown in Figure 7a,b,d,e for both cocurrent and counterflow configurations.

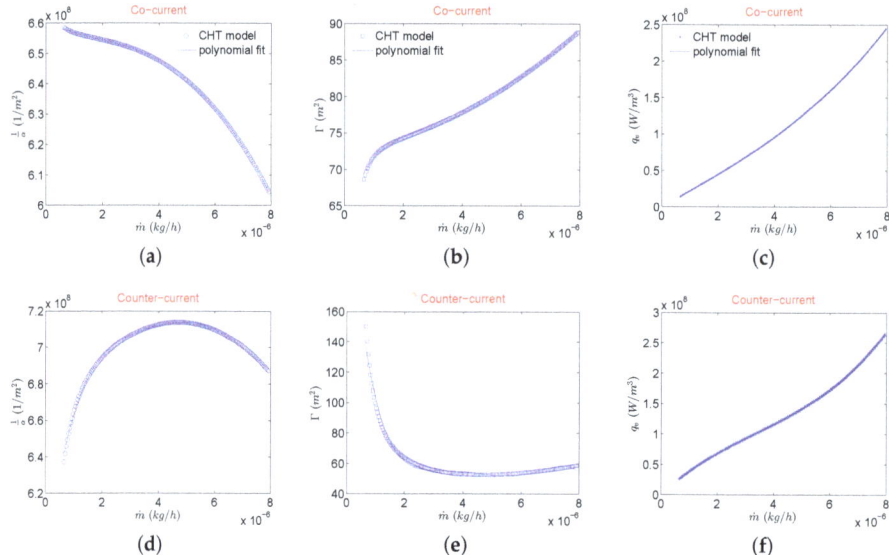

Figure 7. Viscous coefficient (**a,d**), inertial coefficient (**b,e**), and volumetric heat source term (**c,f**) extracted from CHT analysis in both flow configurations.

To model the desired temperature drop on the hot fluid side, a source term q_v is extracted from the hot fluid stream of CHT analysis using Equation (9). The variation of this volumetric heat source term with the mass flow rate is shown in Figures 7c (cocurrent) and 7f (counter flow). A polynomial fit on this q_v is given as a source term to energy equation in the porous model of µHx.

4.2. Porous Model

The presence of manifolds introduces flow maldistribution that affects both pressure drop as well as heat exchanger effectiveness of a µHx compared to earlier performed CHT analysis. A single layer of the µHx as shown in Figure 3 is considered in the porous model. This is due to the fact that pressure and temperature drops of hot side from CHT analyses that are used to extract porous medium coefficients and source term, respectively, already catered for any conjugate heat transfer effects of both layers. Therefore, modeling only one layer with these coefficients and source term should be sufficient to emulate a double layer µHx. It is important to emphasize at this point that simulating only one layer essentially assumes that the thermal performance evaluated from both hot and cold layers of the µHx will be the same, which does not hold valid for an unbalanced (different mass flow rates on the hot and cold sides) µHx. Therefore, for every change in the geometry of the microchannel and/or unbalanced flow conditions, new coefficients (see Figure 7) are to be attained. Moreover, as porous media coefficients as well as the source term are mass flow rate dependent, based upon encountered maldistribution in the manifold, each MC will exhibit respective pressure and temperature drop along the length. Therefore, contrary to the conventional isotropic porous medium approach to model the MC core as in [1,2,5–7], the current porous model behaves as an anisotropic porous medium. This essentially means that each microchannel will exhibit different pressure and temperature drop that enables the observation of the extent of the flow maldistribution.

The experimental pressure drop of the double layer µHx being considered in this study for different flow configurations has been reported by Gerken et al. [14]. Pressure drop showed dependence on the flow configuration as well as the material and thickness of the partition foil employed during the experimental tests. A possible reason for this deviation between different foil materials and thicknesses was associated with possible bending of the thin partition foils inside manifolds although a strong layer of circular pillars was realized underneath (in manifolds) to protect against such undesired deflection of partition foil. Computational results of pressure drop from the current porous model are compared with the experimental results reported for the same µHx from two different studies [13,14] in Figure 8. As results differ from one separation foil material to the other, only results with stainless steel foil with a thickness of 100 µm (as utilized in the current study) are compared and are shown in Figure 8a.

It can be seen that the total pressure drop of the device shows a good agreement between the average of two experimental investigations on the same µHx. Results are more compliant with the results of Yang et al. [13] for smaller mass flows while they match better with Gerken et al. [14] for higher mass flow rates. Pressure drops in the distributing manifold and MC core are also shown in Figure 8b where there exists a very good match between the current porous model and experimental results of Gerken et al. [14]. However, the pressure drop of MC core is slightly overestimated at higher mass flow rates with modeled porous medium coefficients with incorporated source term in the energy equation.

Flow maldistribution is shown in Figure 9 for both cocurrent and counterflow configurations where MC indexing is done as outlined in Figure 3a. As expected, maldistribution shows a weak dependence on the flow configuration where except for the furthest MCs from the inlet, it shows similar patterns in both flow configurations. However, due to the presence of circular pillars, maldistribution does not exhibit a typical profile to be expected of triangular manifolds [9,23]. Another reason for this could be the orthogonal direction of the inlet with reference to the channel flow whereas in most experimental as well as numerical studies, an inline flow is encountered where fluid enters parallel to the base plane of distributing manifold.

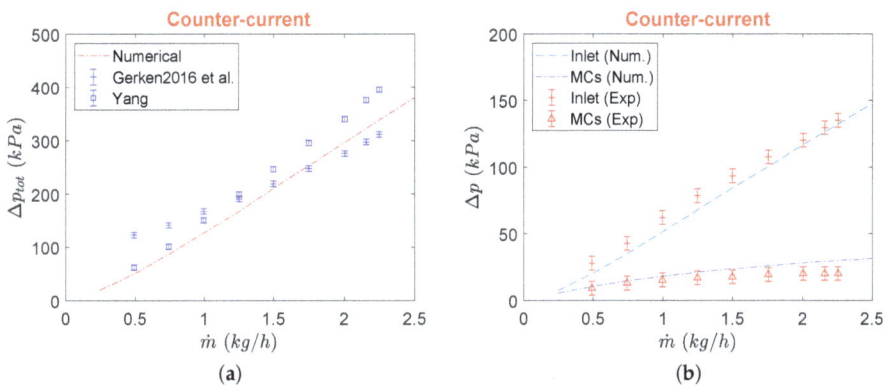

Figure 8. Comparison between experimental and numerical total pressure drop of µHx (**a**), and in the inlet and MCs only (**b**).

To compare the effect of heat transfer on the flow maldistribution, counterflow porous model results are calculated at mass flow rate of 2.5 kg/s by deactivating the energy source term and results are shown in Figure 10.

It is evident that temperature drop does not substantially affect the maldistribution pattern in the middle core of the current manifold while it is higher on either extreme for the case when source term is not modeled. In other words, for current configuration heat transfer helps decreasing the maldistribution in first and last MCs.

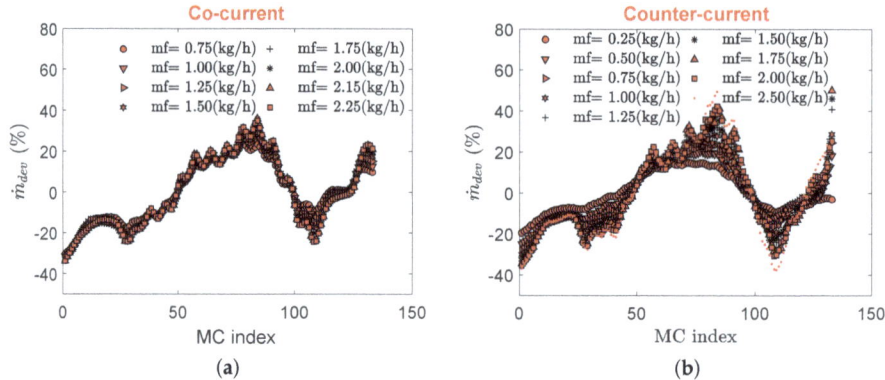

Figure 9. Flow maldistribution in MCs for cocurrent flow (**a**), and counterflow (**b**).

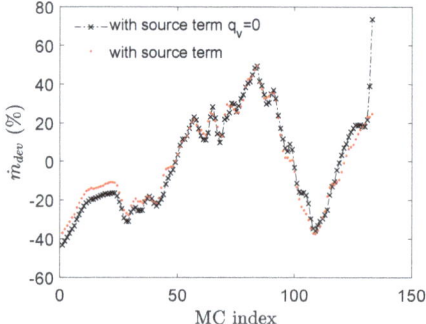

Figure 10. Flow maldistribution with and without source term q_v for counterflow configuration with $\dot{m}_f = 2.5$ kg/s.

Heat exchanger effectiveness evaluated using Equation (14) is shown in Figure 11 for both flow configurations. In the experimental campaign, one pressure and two temperature sensors were placed 0.4 mm from the inlet and outlet of the MCs core to represent an average value of pressure drop and temperature drop or gain through the hot or cold side of MCs core. For a cocurrent configuration, porous model predictions match to the experimental results within the experimental uncertainty. However, for the counterflow configuration, results of the porous model seem to underestimate ε compared to experimental results. To investigate this discrepancy, average fluid temperatures at two planes of 0.5 mm × 0.5 mm are used close to the locations where experimental temperature sensors were placed. The average temperature of the gas is then calculated using a simple equal weighted average of these two temperatures as done in the experimental campaign. As it can be seen in Figure 11b, when temperature estimation at the inlet and outlet of MCs core is done in a way similar to experimental settings, a higher temperature drop and an apparent increase in ε is evidenced. However, when mass flow weighted averages of temperature at inlet and outlet of MC core are considered from numerical simulations, ε is slightly lower than the experimental one. On the other hand, a limited number of sensors are practical limitations inside microdevices therefore, averaging performed using a limited number of temperature and pressure sensors would always result in a slight discrepancy between experimental and numerical global characteristics of the μHxs. Results of ε from the porous model are also compared with the CHT analysis. For counterflow configuration, it can be noted from Figure 11b that porous model in which flow maldistribution is accounted for gives an identical overall heat exchanging efficiency to that obtained with CHT model without manifolds.

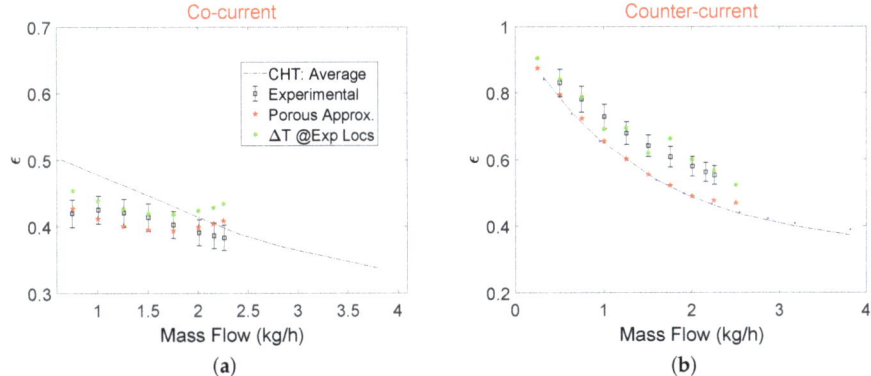

Figure 11. µHx effectiveness for cocurrent flow (**a**), and counterflow (**b**).

A similar conclusion was also presented by Joseph et al. [9] where CHT analysis with periodic side walls for a 60 layered counterflow compact heat exchanger showed only 4% deviation in ε compared to experimental results. This hints that it is sufficient to use a CHT model to predict the thermal effectiveness of a multilayered parallel channel counterflow µHx. Pressure drop analysis of the entire device, however, cannot be performed using only CHT analysis as there is a significant pressure drop in the distributing and collecting manifolds. On the contrary, ε for the cocurrent configuration shows a dependence on the maldistribution, in fact CHT model overestimates ε when compared to porous model (see Figure 11a).

5. Conclusions

Independent of the flow configurations investigated in this study, the developed porous medium-based numerical model shows an excellent match with experimental results for both cocurrent and counterflow configurations. Moreover, as flow inside porous MCs is not treated as wall bounded, rather a free slip boundary condition is applied, a substantial amount of computational power is saved in the process. Since the proposed porous medium model requires only one layer of µHx, with MCs core modeled as porous medium, it eliminates the need of simulating all the layers of such a microdevice where a layered structure is quite common and an effort to computationally model such a device with all the layers would require a staggering amount of computational power. Therefore, the proposed model for the analysis of parallel channel µHx can prove to be a feasible option for rapid design optimization in engineering studies. Based on the discussion conducted earlier, the following conclusions can be inferred:

1. Porous medium coefficients for parallel channel µHx can be extracted for compressible fluids by modifying the existing Darcy–Forchheimer law to incorporate for the strong density variations with increasing mass flow rates in MCs.
2. CHT analysis revealed that gas in both hot and cold fluid streams experiences a "self-cooling" phenomenon where the temperature of the gases keeps on decreasing from inlet to the outlet at higher mass flow rates (Re). Therefore for a µHx operating under balanced mass flow rates, smaller values of mass flow rates are recommended.
3. Pressure drop of the porous model is much higher compared to the CHT due to the presence of collecting and dividing manifolds. Pressure drop estimation using the porous model is in good agreement with the experimental results of the same µHx.
4. Overall heat exchanger effectiveness of a µHx in a counterflow configuration is identical to that of the CHT analysis on the range of mass flows investigated in this study. For a cocurrent

configuration, however, heat exchanger effectiveness from porous model matches well with experimental results while the CHT model overpredicts it.

5. Compared to the meshing strategy adopted in CHT analysis, the porous model results in saving of at least 20 million computational nodes for the double layer gas to gas µHx investigated in this study with good enough predictions of global pressure drop and heat exchanger effectiveness.

Author Contributions: Formal analysis, D.R. & J.J.; Funding acquisition, G.L.M. & M.D. & J.B.; Investigation, Danish Rehman; Methodology, D.R. & J.J. & M.D.; Supervision, G.L.M. & M.D. & J.B.; Validation, D.R. & J.J.; Writing—original draft, D.R.; Writing—review & editing, G.L.M. & J.B. & M.D. All authors have read and agreed to the published version of the manuscript.

Funding: This research received funding from the European Union's Framework Programme for Research and Innovation Horizon 2020 (2014–2020) under the Marie Sklodowska-Curie Grant Agreement No. 643095 (MIGRATE Project).

Conflicts of Interest: The authors declare no conflict of interest.

References

1. Kim, S.J.; Kim, D. Forced Convection in Microstructures for Electronic Equipment Cooling. *ASME J. Heat Transf.* **1999**, *121*, 639–645. [CrossRef]
2. Kim, S.J.; Kim, D.; Lee, D.Y. On the Local Thermal Equilibrium in Microchannel Heat Sinks. *Int. J. Heat Mass Transf.* **2000**, *43*, 1735–1748. [CrossRef]
3. Tuckerman, D.; Pease, R. High-performance heat sinking for VLSI. *IEEE Electron Device Lett.* **1981**, *5*, 126–129. [CrossRef]
4. Knight, R.; Goodling, J.; Hall, D. Optimal Thermal Design of Forced Convection Heat Sinks-Analytical. *ASME J. Electron. Packag.* **1991**, *113*, 313–321. [CrossRef]
5. Dong, L.; Garimella, S. Analysis and Optimization of the Termal Performance of Microchannel Heat Sinks. *Int. J. Numer. Heat Fluid Flow* **2005**, *15*, 7–26.
6. Lim, F.Y.; Abdullah, S.; Ahmad, I. Numerical Study of Fluid Flow and Heat Transfer in MicroChannel Heat Sinks using Anisotropic Porous Media Approximation. *J. Appl. Sci.* **2010**, *10*, 2047–2057. [CrossRef]
7. Alm, B.; Imke, U.; Knitter, R.; Schygulla, U.; Zimmermann, S. Testing and simulation of ceramic micro heat exchangers. *Chem. Eng. J.* **2008**, *135*, S179–S184. [CrossRef]
8. Joseph, J.; Nacereddine, R.; Delanaye, M.; Giraldo, A.; Roubah, M.; Brandner, J. Advanced CFD methodology to investigate heat exchanger performance and experimental validation of a reduced model for a micro-CHP application. In Proceedings of the 6th Micro and Nano Flows Conference (MNF2018), Atlanta, GA, USA, 6–7 September 2018.
9. Joseph, J.; Nacereddine, R.; Delanaye, M.; Korvink, J.G.; Brandner, J.J. Advanced Numerical Methodology to Analyze High-Temperature Wire-Net Compact Heat Exchangers For a Micro-Combined Heat and Power System Application. *Heat Transf. Eng.* **2019**, *41*, 1–13. [CrossRef]
10. Joseph, J.; Delanaye, M.; Nacereddine, R.; Giraldo, A.; Rouabah, M.; Korvink, J.G.; Brandner, J.J. Numerical Study of Perturbators Influence on Heat Transfer and Investigation of Collector Performance for a Micro-Combined Heat and Power System Application. *Heat Transf. Eng.* **2020**, *42*, 1–23. [CrossRef]
11. Rehman, D.; Joseph, J.; Morini, G.; Delanaye, M.; Brandner, J. A Porous Media Model for a Double Layered Gas-to-Gas Micro Heat Exchanger operating in Laminar Flow Regime. In Proceedings of the 37th UIT Heat Transfer Conference, Padova, Italy, 24–26 June 2019.
12. Yang, Y.; Morini, G.L.; Brandner, J.J. Experimental analysis of the influence of wall axial conduction on gas-to-gas micro heat exchanger effectiveness. *Int. J. Heat Mass Transf.* **2014**, *69*, 17–25. [CrossRef]
13. Yang, Y.; Gerken, I.; Brandner, J.J.; Morini, G.L. Design and Experimental Investigation of a Gas-to-Gas Counter-Flow Micro Heat Exchanger. *Exp. Heat Transf.* **2014**, *27*, 340–359. [CrossRef]
14. Gerken, I.; Brandner, J.J.; Dittmeyer, R. Heat transfer enhancement with gas-to-gas micro heat exchangers. *Appl. Therm. Eng.* **2016**, *93*, 1410–1416. [CrossRef]
15. Yang, Y. Experimental and Numerical Analysis of Gas Forced Convection through Microtubes and Micro Heat Exchangers. Ph.D. Dissertation, Universita di Bologna, Bologna, Italy, 2012.

16. Abraham, J.; Sparrow, E.; Tong, J. Breakdown of laminar pipe flow into transitional intermittency and subsequent attainment of fully developed intermittent or turbulent flow. *Numer. Heat Transf. Part B Fundam.* **2008**, *54*, 103–115. [CrossRef]
17. Abraham, J.; Sparrow, E.; Minkowycz, W. Internal-flow Nusselt numbers for the low-Reynolds-number end of the laminar-to-turbulent transition regime. *Int. J. Heat Mass Transf.* **2011**, *54*, 584–588. [CrossRef]
18. Minkowycz, W.; Abraham, J.; Sparrow, E. Numerical simulation of laminar breakdown and subsequent intermittent and turbulent flow in parallel-plate channels: Effects of inlet velocity profile and turbulence intensity. *Int. J. Heat Mass Transf.* **2009**, *52*, 4040–4046. [CrossRef]
19. Koyama, K.; Asako, Y. Experimental Investigation of Heat Transfer Characteristics on a Gas-to-Gas Parallel Flow Microchannel Heat Exchanger. *Open Transp. Phenom. J.* **2010**, *2*, 1–8. [CrossRef]
20. Koyama, K.; Asako, Y. Experimental Investigation of Heat Transfer Characteristics on a Gas-to-Gas Counterflow Microchannel Heat Exchanger. *Exp. Heat Transf.* **2010**, *23*, 130–143. [CrossRef]
21. Rehman, D.; Morini, G.L.; Hong, C. A Comparison of Data Reduction Methods for Average Friction Factor Calculation of Adiabatic Gas Flows in Microchannels. *Micromachines* **2019**, *10*, 171. [CrossRef] [PubMed]
22. Hong, C.; Tanaka, G.; Asako, Y.; Katanoda, H. Flow characteristics of gaseous flow through a microtube discharged into the atmosphere. *Int. J. Heat Mass Transf.* **2018**, *121*, 187–195. [CrossRef]
23. Renault, C.; Colin, S.; Orieux, S.; Cognet, P.; Tzédakis, T. Optimal design of multi-channel microreactor for uniform residence time distribution. *Microsyst. Technol.* **2012**, *18*, 209–223. [CrossRef]

© 2020 by the authors. Licensee MDPI, Basel, Switzerland. This article is an open access article distributed under the terms and conditions of the Creative Commons Attribution (CC BY) license (http://creativecommons.org/licenses/by/4.0/).

Article

Numerical and Experimental Study of Microchannel Performance on Flow Maldistribution

Jojomon Joseph [1,2,*], Danish Rehman [3], Michel Delanaye [1], Gian Luca Morini [3], Rabia Nacereddine [1], Jan G. Korvink [2] and Juergen J. Brandner [2]

1. MITIS SA, Rue del Rodje Cinse 98, 4102 Seraing, Belgium; michel.delanaye@mitis.be (M.D.); rabia.nacereddine@mitis.be (R.N.)
2. Institute of Microstructure Technology, Karlsruhe Institute for Technology, 76131 Karlsruhe, Germany; jan.korvink@kit.edu (J.G.K.); juergen.brandner@kit.edu (J.J.B.)
3. Microfluidics Laboratory, Department of Industrial Engineering (DIN), University of Bologna, Via del Lazzaretto 15/5, 40131 Bologna BO, Italy; danish.rehman2@unibo.it (D.R.); gianluca.morini3@unibo.it (G.L.M.)
* Correspondence: joseph.jojomon@mitis.be

Received: 27 February 2020; Accepted: 19 March 2020; Published: 20 March 2020

Abstract: Miniaturized heat exchangers are well known for their superior heat transfer capabilities in comparison to macro-scale devices. While in standard microchannel systems the improved performance is provided by miniaturized distances and very small hydraulic diameters, another approach can also be followed, namely, the generation of local turbulences. Localized turbulence enhances the heat exchanger performance in any channel or tube, but also includes an increased pressure loss. Shifting the critical Reynolds number to a lower value by introducing perturbators controls pressure losses and improves thermal efficiency to a considerable extent. The objective of this paper is to investigate in detail collector performance based on reduced-order modelling and validate the numerical model based on experimental observations of flow maldistribution and pressure losses. Two different types of perturbators, Wire-net and S-shape, were analyzed. For the former, a metallic wire mesh was inserted in the flow passages (hot and cold gas flow) to ensure stiffness and enhance microchannel efficiency. The wire-net perturbators were replaced using an S-shaped perturbator model for a comparative study in the second case mentioned above. An optimum mass flow rate could be found when the thermal efficiency reaches a maximum. Investigation of collectors with different microchannel configurations (s-shaped, wire-net and plane channels) showed that mass flow rate deviation decreases with an increase in microchannel resistance. The recirculation zones in the cylindrical collectors also changed the maldistribution pattern. From experiments, it could be observed that microchannels with S-shaped perturbators shifted the onset of turbulent transition to lower Reynolds number values. Experimental studies on pressure losses showed that the pressure losses obtained from numerical studies were in good agreement with the experiments (<4%).

Keywords: micro channel; reduced model; wire-net perturbators; s-shaped perturbators; high-temperature heat exchangers

---

1. Introduction

In recent years, Computational Fluid Dynamics (CFD) has been widely used to analyze heat exchanger performance and optimize it for specific applications. In the micro-gas turbine (MGT) industry, it is challenging to maintain laminar flow in microchannels. On the other hand, it is well-known that perturbators lead to local turbulences and, therefore, increase the heat transfer performance. Min et al. [1] presented different heat exchanger design schemes for gas turbine applications. They suggested that heat transfer effectiveness must be higher than 90% and pressure losses must be less than 3%, along with excellent resistance to oxidation and creep for temperatures above 650 °C. Based on a recent market study on MGT for decentralized energy systems made by Xiao [2], the target for thermal effectiveness must be higher than 90% and pressure losses must be less than 5%. Shah [3], in his review paper, noted similar performance requirements.

Microchannel entrance length effects, as well as flow maldistribution [4], plays an essential role in compact heat exchanger performance. The temperature distribution effect on the flow maldistribution for a parallel microchannel system was meticulously investigated by Siva [5]. He figured out that a high heat flux or highly heated areas induced a reduction in viscosity of fluid, resulting in a higher flow maldistribution. Different types of collectors were analyzed by Siddique [6], and it was found that the creation of a stagnation zone, the growth of a boundary layer along the collector wall and low/high-velocity zones in the collector were the prime causes of flow maldistribution.

Microchannel performance plays a vital role in heat exchanger assembly. Tao [7] suggested heat transfer enhancement by decreasing the thermal boundary layer, introducing protrusions inside the microchannels and increasing the velocity gradient near the heated surface. Transverse vortex generators and longitudinal vortex generators play a significant role in local and global heat transfer enhancements, as shown in [8]. Valencia [9] showed that at Reynolds number 2000 heat transfer was enhanced by 30% due to turbulence together with a five-fold increase in friction factor. Increasing the blockage ratio and reducing the aspect ratio increased the global Nusselt number. An analysis by Paolo [10] of Conjugate Heat Transfer (CHT) on a cube revealed that the heat transfer was minimum when the boundary layer was undisturbed.

Smooth pipes (with smaller roughness) have friction factor behavior similar to the conventional theory. On the contrary, rough tubes have showed an earlier turbulent transition (Re = 350) [11–14]. Thus, adding perturbators in microchannels can induce turbulent transitions at smaller Reynolds numbers. At the same time, near-wall turbulence causes a significant share of pressure losses. Increasing the area density will increase the pressure losses and result in an increased fouling effect. Studies have showed that a high level of turbulence intensifies the self-cleaning inside complex microchannels [15]. On the contrary, increasing size to overcome pressure losses results in higher capital and reduces the performance advantages of miniaturization.

Yang [16,17] compared the performance of heat exchange devices with different microstructures manufactured on thin foils to enhance heat transfer. He found that "partition foils having low thermal conductivity can enhance the heat transfer by decreasing the axial conduction" and that, at a microscale, cross-flow configuration tends to provide similar results to the counter-current arrangement at a large Reynolds number (>2400). This is because, even if the arrangement is counter-current, there is always a cross-flow part in microchannel arrangements due to the microchannels with plane partition foils [17]. This fact led to the wire-net arrangement in our study, which was created to suppress the cross-flow pattern in the counter-flow channels.

Microchannel performances of wire-net and S-shape perturbators were presented by Joseph [18]. Joseph [18–21] also proposed a Reduced Order Model (ROM) to investigate the collector performance. The global parameters of the heat exchanger (overall thermal efficiency and pressure losses) were investigated using the ROM and were validated experimentally based on high-temperature testing of the full-scale heat exchanger. The experimental study used to validate the employed numerical model in terms of flow maldistribution, microchannel pressure loss and so on was not presented. In this paper, an experimental study is performed to investigate flow maldistribution and microchannel pressure losses based on a pressure-loss study (without temperature effects). In addition, an experimentally validated ROM was utilized to investigate the influence of collector performance on microchannel characteristics like foil thickness, type of perturbators and collector recirculation zones.

2. Numerical Model

A single heat exchanger block comprises of a secondary collector with 60 microchannels. The number of blocks is calculated based on the total mass flow rate, required pressure losses and thermal efficiency. A primary collector supplies and collects the mass flow rate to each block of heat exchangers. Thus, flow maldistribution in both primary and secondary collectors play a crucial role in the performance of the heat exchanger assembly. A computationally inexpensive methodology that has been presented by Joseph [20] will be utilized for a detailed investigation of collector performance.

2.1. Microchannel Design

Two types of microchannel configurations, wire-net and S-shape perturbators, were selected for detailed microchannel CFD analysis. A comparative study was carried out by replacing the wire-net perturbators with S-shape perturbators. The wire-net heat exchanger consist of partition foils that are separated by bobbing wire-net structures. The patented wire-net heat exchanger [18–21] is assembled as a stack of counter-flow channels with optimized thicknesses separated by thin partition foils. The frames are brazed together with the partition foils (see Figure 1a). Cylindrical collectors provide cylindrical-shaped microchannel inlets that distribute mass flow rates uniformly to the wire-net microchannels. Cylindrical collectors with wire-net microchannels are depicted in Figure 1a.

High computational effort is needed to simulate entire counter-flow wire-net microchannels. Two types of geometrical symmetry exist in a plane that is parallel to the inlet flow direction (see Figure 1a). The primary symmetrical domain (see Figure 1b) is at half of the secondary collector's diameter, while the secondary symmetrical domain is located at the center of the wire-net compact heat exchanger (see Figure 1c). Again, high computational effort is needed to carry out a Conjugate Heat Transfer (CHT) analysis by holding the primary symmetry (see Figure 1b). The objective of the preliminary analysis based on the primary symmetrical domain is to investigate the homogeneity in the wire-net flow physics and verify the second symmetry assumption for the CHT analysis. Thus, a turbulent K-ω Reynolds averaged Navier-Stokes (NS) model without energy equations was adopted for the preliminary study to reduce solver complexity. Apparently, the second analysis based on the second symmetrical domain was performed to investigate the Conjugate Heat Transfer effects in microchannels.

Similarly, periodic boundary conditions were implemented in the plane that is perpendicular to the inlet flow direction. A periodicity exists in half of the microchannel foil thickness on every hot and cold side alternatively. Thus, the counter-flow arrangement was simplified into single hot and cold microchannels with partition foil above, as shown in Figure 1c.

Three-dimensional, ideal gas, steady Conjugate Heat Transfer (CHT) simulations were carried out in counter-flow ducts to investigate the effect of thermal efficiency and pressure losses for various inlet mass flow rates. A turbulent K-ω shear stress transport model was utilized for all the CHT models and collectors simulations. The free microchannel simulations were performed using laminar models. The k-ε model was not accurate in the near-wall region, while the K-ω model was appropriate for near-wall turbulent flows. In the K-ω shear-stress turbulent transport model, both conventional models (K-ω & k-ε) are mixed to take advantage of their unique advantages [22].

The microchannel thermal efficiency ε [%] is calculated using the following:

$$\varepsilon [\%] = \frac{C_h(T_{in}^h - T_{out}^h)}{min(C_c, C_h)(T_{in}^h - T_{in}^c)} \times 100 \qquad (1)$$

$$C_c = m_{\mu c} Cp_c \; ; C_h = m_{\mu h} Cp_h \qquad (2)$$

where $Cp_{c/h}$ are the hot and cold fluid capacity rates, respectively; $T_{in/out}^h$ and $T_{in/out}^c$ are the inlet and outlet temperatures of the counter-flow channels, respectively; and $m_{\mu c/\mu h}$ are the microchannel mass flow rates at the hot and cold inlets, respectively. The microchannel pressure losses ΔP [%] were calculated using the following:

$$\Delta P [\%] = \frac{P_{in} - P_{out}}{P_{ab}} \times 100 \qquad (3)$$

where $P_{in/out}$ are the inlet and outlet absolute pressure, respectively; and P_{ab} is the absolute pressure at the inlet. The influence of pressure losses on density is relatively low if the total pressure loss is less than 5%. Morini [23] considered in his studies that flow is only incompressible when the Mach number is less than 0.3, or when the relative pressure loss $\Delta P/P_{ab}$ is less than 5%. In our present study, the maximum pressure loss expected was 5%. As a result, the flow regime remains incompressible, although density variations as a function of the inverse of temperature as well as the fluid were dilatable [23]. As a result, a segregated solver is adopted.

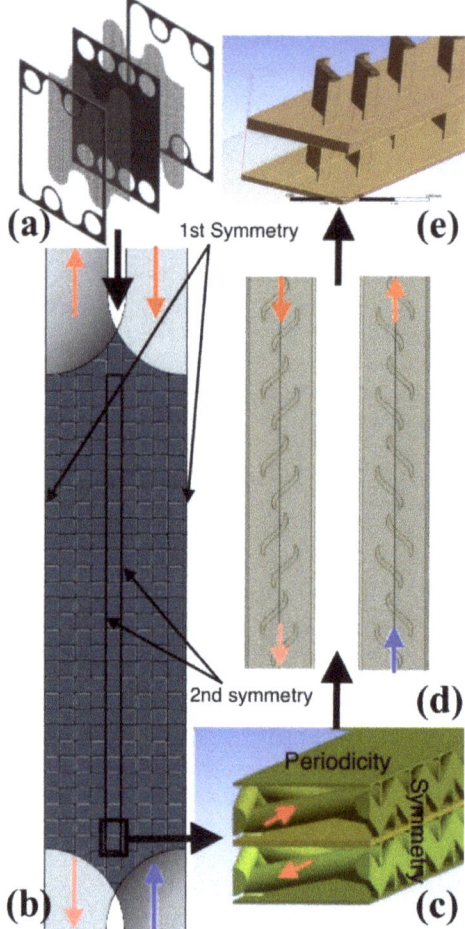

Figure 1. Wire-net brazed heat exchanger arrangement (**a**); microchannels with wire-net perturbators (**b**); Conjugate Heat Transfer (CHT) domain (**c**); wire-net microchannels replaced by S-shaped perturbators (**d**); CHT domain (**e**).

2.2. Porous Medium Model

The CFD methodology based on a porous medium approximation was implemented to investigate the flow maldistribution. A general scheme of the heat exchanger arrangement is depicted in Figure 2a. The microchannels of the wire-net/S-shape fins have a complicated structure that again leads to high simulation needs. The best method is to split the heat exchanger configuration part by part and analyze the performance using a Reduced Order Model (ROM).

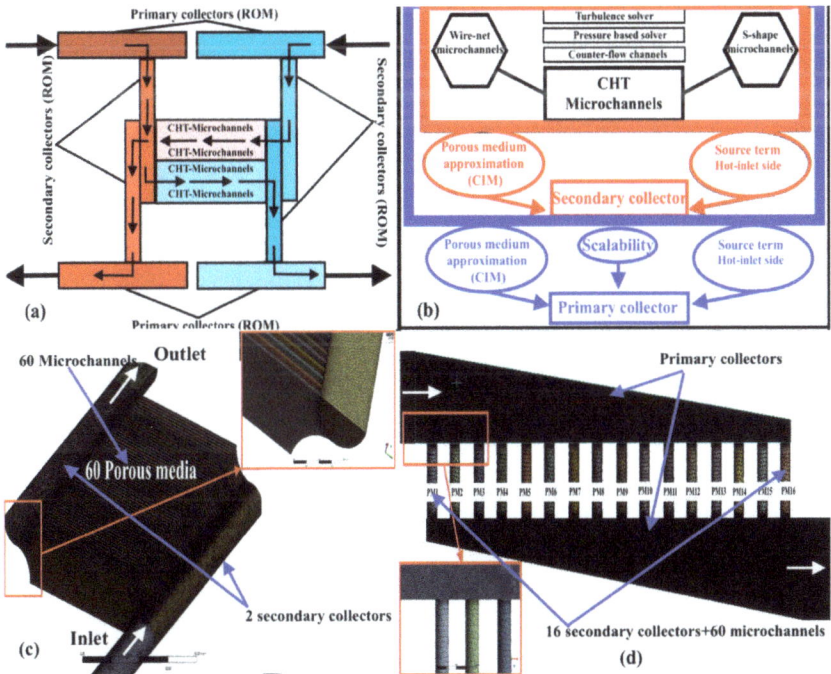

Figure 2. Scheme of the heat exchanger assembly (**a**); numerical methodology (**b**); secondary collector model (**c**); primary collector mesh model with 16 porous mediums replicating 16 collectors (**d**).

The modelling was performed in three different stages (see Figure 2b):

1. Microchannel CHT model. Estimating the microchannel performance based on detailed three-dimensional CHT analysis.
2. Reduced model for cylindrical secondary collectors (see Figure 2c). Microchannels are replaced by porous medium models to evaluate the flow maldistribution and thereby the secondary collector performance. The microchannels have a very coarse mesh (10,000 cells per channel), and the secondary collector has a highly refined mesh to capture the secondary effects.
3. Reduced model for trapezoidal primary collectors (see Figure 2d). Scalability of the ROM was utilized to model both the microchannel along with the secondary collectors as a single porous medium. Therefore, the effects of both microchannels and secondary collectors were introduced into a cylindrical duct. The cylindrical duct has a very coarse mesh, and the primary collector has a highly refined mesh. An overall heat exchanger performance is estimated based on flow maldistribution and overall pressure losses.

In the first stage of the CFD modelling, a detailed CHT analysis was carried out for a microchannel section with specific boundary conditions. Microchannel performance characteristics were utilized to model the microchannels as a porous medium, with inertial and viscous coefficients as the main parameters.

In the second stage of modelling, the microchannels were considered a porous medium, and the cylindrical secondary collector performance was evaluated. The cylindrical cross-section of the secondary collectors was extruded at both the outlet and inlet for better convergence, as depicted in

Figure 2c. The Constant Integration Method (CIM) based on the Darcy-Forchheimer law was utilized to calculate the porous medium characteristics that replicate the Conjugate Heat Transfer model. The computational domain was sharply reduced, since the microchannels were examined as a porous medium with a free-slip boundary condition.

In the third stage of modelling, the microchannels with secondary collectors were considered a porous medium, and the scalability of the ROM provided an easiness with which to create the temperature and pressure jump across circular of cross-sections. Scalability of the ROM was developed to characterize the primary collector performance and thereby the overall heat exchanger performance for the micro combined heat and power (CHP) system.

The primary and secondary collector mesh model is depicted in Figure 2a,b, respectively. Half a section of the cylindrical collectors that feeds 60 microchannels is depicted in Figure 2a. The global performance of a secondary collector with 60 microchannels was implemented into a cylindrical duct with a trapezoidal primary collector configuration (see Figure 2b). A cylindrical channel with a porous medium approximation replaced the effects of the secondary collector and 60 microchannels.

The number of microchannels, together with the microchannel size reduction (γ) coefficient, was utilized to calculate the microchannel inlet mass flow rate from the collector inlet mass flow rate,

$$m_{\mu CHT} = \frac{m_{co}}{\gamma n_{pl}} \quad (4)$$

where $m_{\mu CHT}$ is the microchannel mass flow rate, m_{co} is the collector mass flow rate, n_{pl} is the number of plates/channels and γ is the microchannel size-reduction coefficient calculated due to simplification of the computational domain (symmetric and periodic boundary conditions). The primary collector inlet mass flow rate was calculated using the following:

$$m_{co} = \frac{m_{to}}{2 n_{bl}} \quad (5)$$

where m_{co} is the secondary collector mass flow rate, m_{to} is the total mass flow rate and n_{bl} is the number of blocks/modules.

Reduced Order Model (ROM)

Global parameters such as pressure loss, thermal efficiency and turbulent viscosity were determined from a three-dimensional CHT analysis of microchannels. This was utilized to calculate the inertial and viscous coefficients (porous medium model) of the reduced model using CIM. The Darcy-Forchheimer law was modified and implemented to account for the significant temperature variation (T = 600 K) and localized turbulence effect in the microchannels [19]:

$$\begin{aligned}\frac{(P_{in}-P_{out})}{L} &= \frac{1}{\alpha}\frac{L}{\Delta T}\frac{m}{A}\frac{R}{P}\left\{\left[T^{\frac{1}{2}}\left(3T^2 - 5S_1 T + 15S_1^2\right) - 15S_1^{\frac{5}{2}}\arctan\left(\frac{T}{S_1}\right)^{\frac{1}{2}}\right) + \epsilon_m \frac{PT}{R}\right]_{T_0}^{T_1} \\ &+ \frac{1}{2}C_2 \frac{m}{A}\left[\frac{T^2}{2}\right]_{T_0}^{T_1}\right\}\end{aligned} \quad (6)$$

Power per unit volume was implemented as a source term along the microchannels (see Equation (7)). Joseph [19,20] applied the new approach and validated the porous medium model (with modified inertial and viscous coefficients) for secondary collectors.

$$Q = m_{\mu CHT}(T_{in}^{h/c} - T_{out}^{h/c})Cp_{h/c} \quad (7)$$

The model was validated numerically and experimentally [18–20]. Rehman et al. [24] used a similar approach for microchannels with large density variations (compressible flows). The collector's flow maldistribution influenced the mass flow rate and thereby the heat transfer prediction was biased along with total pressure losses.

3. Results and Discussion

3.1. Microchannel Performance

The non-dimensional velocity contours (with and without the wire-net) from the preliminary analysis based on the primary symmetrical domain are depicted in Figure 3. The non-uniform velocity distribution in the microchannels (without the wire-net) adversely affect the compact heat exchanger performance, as mentioned by Yang et al. [25], due to lower resistance from the inlet to the outlet. This reduced the benefits of counter-flow passages, since the maximum velocity gradients lied at two extreme inlets (see Figure 3). However, the microchannels with the wire-net showed a more homogeneous velocity distribution that retained the counter-flow effects (see Figure 3a). These effects were similar to the conventional grid turbulence theory [26], where a grid generates turbulence that is nearly homogeneous and isotropic. Apparently, conventional grid turbulence theory is the major assumption in most of the turbulence models [27]. Thus, the numerical model can be simplified (see Figure 1c,e) for a detailed CHT analysis.

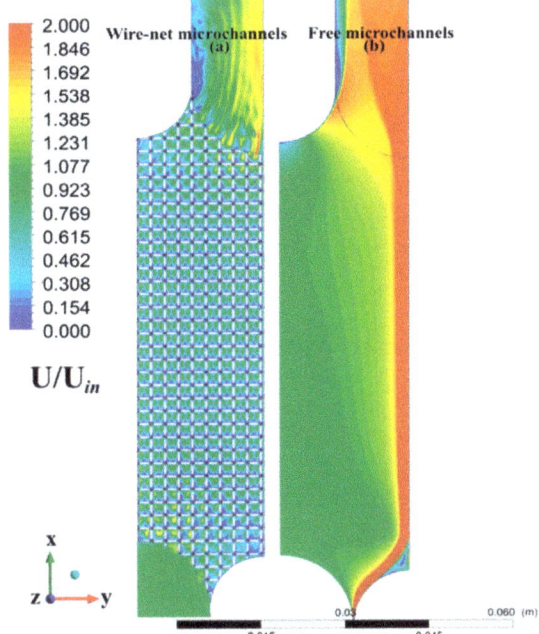

Figure 3. Non-dimensional velocity contours of microchannels with and without the wire-net.

CHT analysis was performed for counter-flow microchannels with symmetric boundary conditions on both sides and periodic boundary conditions of the partition foils. The acceptable thickness for metal 3D-printing is 0.25 mm. Thus, a 0.25 mm thick partition foil was selected for microchannels with S-shaped perturbators. Obviously, there is an optimum mass flow rate where the thermal efficiency reaches a maximum [19]. However, a substantial decrease in efficiency at higher mass flow rates was encountered. The steepness of this efficiency curve pattern was stronger for the S-shape perturbator than the wire-net protrusions (see Figure 4). This steepness increases the working range and thereby reduces the size and cost of the compact heat exchanger. As the steepness decreases, the microchannels can work at high mass flow rates with higher efficiency and, thereby, the number of microchannels required will be lower. As a result, heat exchangers became more compact and cost-effective. However, the pressure loss was relatively high for the wire-net at higher Reynolds numbers. The varying pressure-loss pattern (along with an increasing mass flow rate) was quadratic, which is similar to the conventional Darcy-Forchheimer law where the inertial and viscous terms are well-balanced. The free microchannels had the smallest pressure losses, as expected. The wire-net introduced higher pressure losses than the S-shaped fins (see Figure 4).

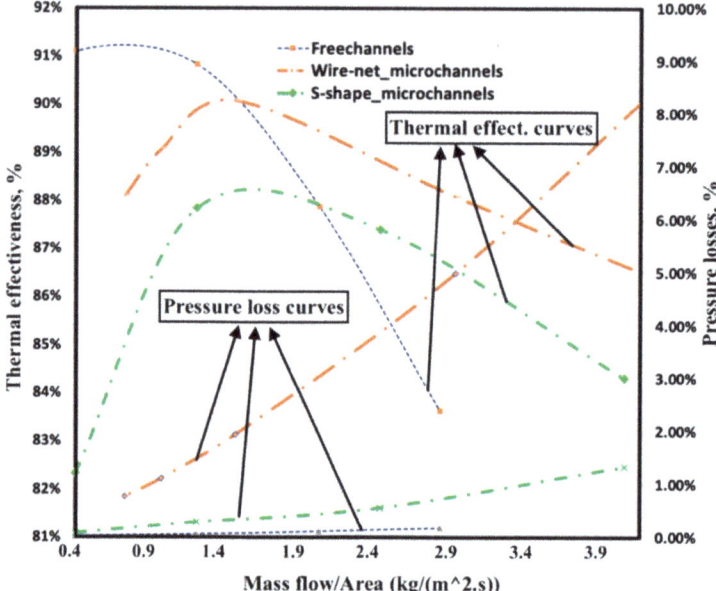

Figure 4. Thermal efficiency and pressure losses of free channels with the wire-net and S-shape microchannels for various mass flow rates based on CHT simulations.

3.2. Collector Performance

3.2.1. Secondary Collector

Influence on Perturbators

The secondary collectors distributed the flow to the 60 microchannels, and the flow maldistribution is depicted in Figure 5. The collectors with lower microchannel resistance (channels with S-shape perturbators) showed a maximum deviation from the ideal CHT mass flow rate. This coincides with the studies made by Thansekhar [28] and Teng [17,29,30], which showed that the higher the flow resistance through the channels, the better the flow distribution, From these studies it can be observed that flow distribution becomes more uniform at higher flow rates, and that flow resistance increases with an increased mass flow rate.

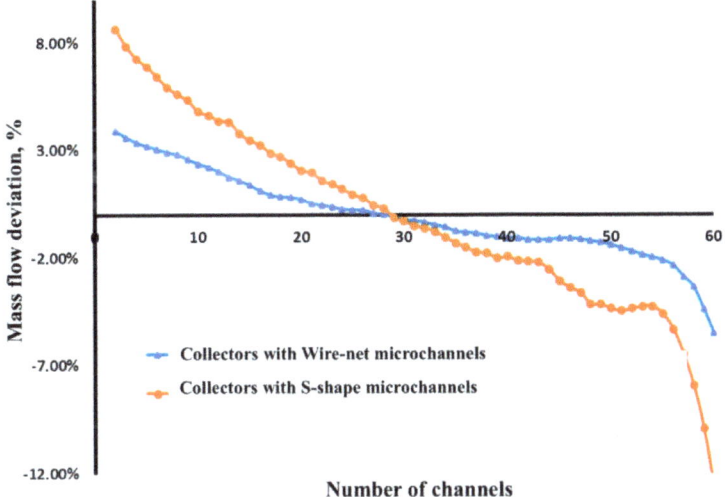

Figure 5. Mass flow rate deviation of perturbators based on Reduced Order Model (ROM).

Influence on Recirculation Zones

Three different configurations were selected to investigate the collector performance:

1. microchannels without the wire-net, $m_{co} = 2.76 \times 10^{-6}$ kg/s;
2. microchannels with the wire-net, $m_{co} = 2.76 \times 10^{-6}$ kg/s; and
3. microchannels with the wire-net, $m_{co} = 3.86 \times 10^{-6}$ kg/s.

The first two configurations were selected to investigate and compare the recirculation zone of cylindrical secondary collectors for different microchannel pressure losses. Further, the second and third configurations were selected to study the effect on collector inlet mass flow rates. Figure 6a shows the vorticity strength (calculated based on the Lambda 2 velocity criteria) of the recirculation zone for three different mass flow rates. The vortices originated at the beginning of the recirculation zone and remained steady enough to dissipate the integral vortex near the collector beds.

Due to the intense mixing of the recirculating fluid with the entraining fluid, strong vortices originated along the recirculation zone. The intensity of the recirculation zones of the cylindrical collector was clear from the vorticity contour (see Figure 6a, left). Thus, the mass flow rate had a strong effect on flow maldistribution, as stated above. As microchannel pressure losses increased (see Figure 6a, center and right), vorticity strength increased. As a result, the mass flow rate distribution near the collector beds became irregular, and deviation from the CHT mass flow rate increased. In addition, the vorticity strength increased with increases in the collector inlet mass flow rate.

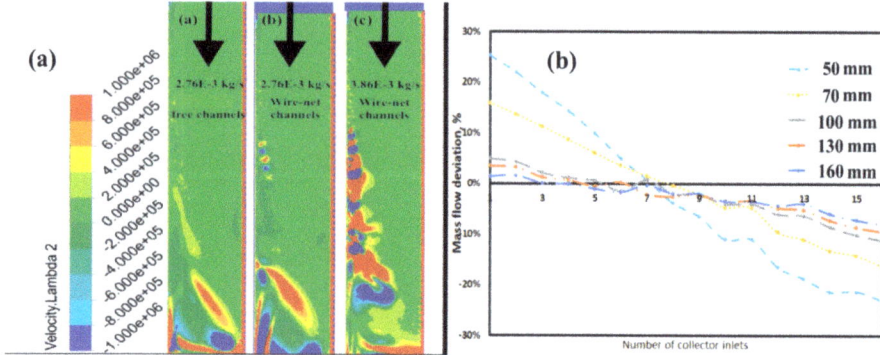

Figure 6. Cylindrical secondary collector performance for various microchannel pressure losses and mass flow rates (**a**); primary collector performance for different heights (**b**) based on a computational study.

3.2.2. Primary Collector

The primary trapezoidal collector inlet distributed the flow to 16 secondary collectors and then to 60 microchannels. Collector flow maldistribution strongly depended on the secondary flows in the primary collectors. A parametric study of different primary collector heights was performed, and the flow maldistribution is depicted in Figure 6b. As the collector height increased, the maldistribution decreased and the optimum collector height was found to be at 100 mm. The velocity contours of the minimum and maximum collector heights are represented in Figure 7a,b, respectively. As the height increased, the velocity near the secondary collector inlet decreased, as did the recirculation strength. The highest turbulence production for the maximum height configuration was near the primary collector outlet. The primary collector height configuration profoundly influenced the flow distribution as well as the turbulence characteristics.

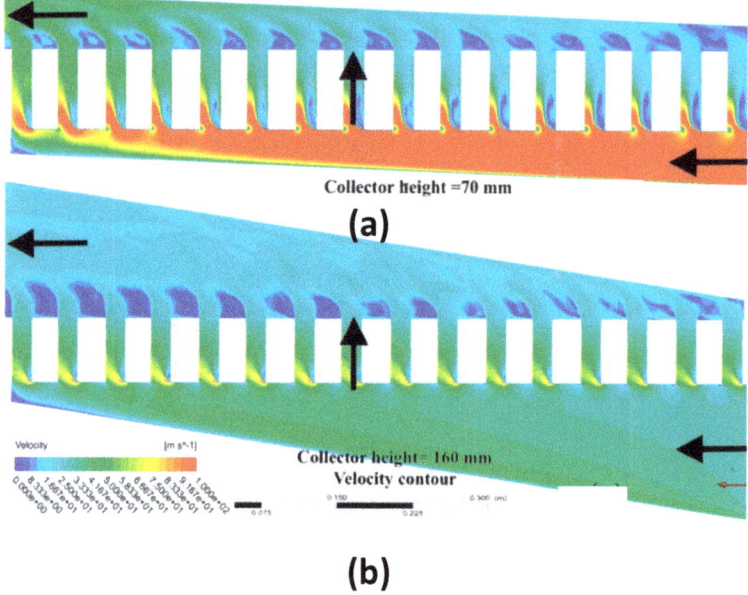

Figure 7. Velocity contour height = 50 mm (**a**); velocity contour height = 160 mm (**b**).

4. Experimental Studies

Using numerical modeling it has been established that the generation of local turbulence serves as the prime reason for enhanced heat transfer when the perturbators are used within microchannels. Such enhanced turbulence in the flow increases the friction factor, but fortunately this increase in the friction factor can easily be observed using an experimental pressure-loss analysis of adiabatic flow, without the need of a high-temperature gas flow. The thermal performance of the micro heat exchanger was numerically analyzed above using two different perturbators. Of those, this section details a microchannel with an S-shaped perturbator that puts into evidence the increased frictional losses and an earlier onset of laminar to turbulent flow transition. A complete CFD investigation of the experimental microchannel is also performed in order to validate the earlier usage of the k-ω SST turbulence model during CHT analysis to model the local turbulent flow around the perturbator. Similarly, usage of the same turbulence model in the collector studies in which microchannels are modeled as porous media can also be justified through a comparison of the experimental and numerical maldistribution patterns of the adiabatic flow. It is acknowledged that the extent of maldistribution in a diabatic operational condition would differ from that of an adiabatic case, but the selection of turbulence model can be well justified using the adiabatic case only. Therefore, trapezoidal collectors with two layers of 33 microchannels were tested experimentally using an adiabatic flow to demonstrate the close agreement between experimental and numerical evaluations of maldistribution. The details of the fabrication, test bench and experimental data reduction, along with the results of both experimental studies, are discussed in this section.

4.1. Microchannel

The test bench and apparatus for microchannels with S-shaped perturbators is depicted in Figure 8. Microchannels with S-shaped perturbators were microfabricated using a micro-milling process as shown in Figure 8b. After passing through a particle filter (2), gas stored in a high-pressure tank (1) was directed to the desired mass flow rate controller using a three way valve (3). A mass flow rate controller (4) with an operating range of 0–5000 NmL/min controlled the inlet flow. Gas was then allowed to enter the microchannel test assembly (5) and finally exit through an outer port. The total pressure loss between the inlet and the outlet of the microchannel assembly was measured utilizing a differential pressure transducer (Validyne DP15) (6) with an interchangeable sensing element that allowed accurate measurements over the whole range of encountered pressures. Atmospheric pressure was measured using an absolute pressure sensor (Validyne AP42) (7). The pressure sensor calibration was performed using a comparison test pump (Giussani BT 400) that was capable of imposing the desired pressure difference relative to the atmosphere.

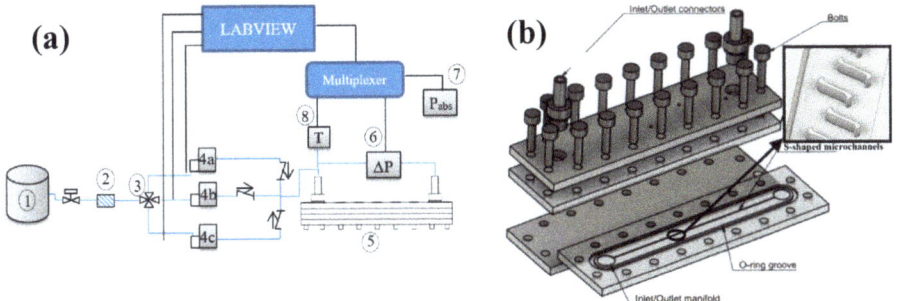

Figure 8. Experimental test bench (**a**) and microchannel assembly (**b**).

To measure the temperature at the entrance of the microchannel, a calibrated K-type thermocouple (8) was used. The thermocouple voltage and an amplified voltage of pressure sensors were fed to the internal multiplexer board of Agilent 39470A and were read through a PC using a Labview program. Uncertainties associated with all the measuring instruments are tabulated in Table 1.

Table 1. Characteristics and uncertainties of the measurement instruments.

Label of Figure 8	Instrument	Range	Uncertainty
(4a)	Flowmeter (Bronkhorst EL-Flow E7000)	0–5000 (nmL/min)	± 0.6% FS
(4b)	Flowmeter (Bronkhorst EL-Flow E7000)	0–500 (nmL/min)	± 0.5% FS
(4c)	Flowmeter (Bronkhorst EL-Flow E7000)	0–50 (nmL/min)	± 0.5% FS
(6)	Differential pressure transducer(Validyne DP15)	0–35 (kPa) 0–86 (kPa) 0–220 (kPa)	± 0.5% FS
(8)	Thermocouple (K type)	0–200 (°C)	± 0.2 K

The experimental friction factor was established using the following [31]:

$$f_f = \left(\frac{D_h}{L}\right)\left[\frac{P_{in}^2 - P_{out}^2}{RT_{av}\left(\frac{m}{\Omega}\right)^2} - 2\ln\left(\frac{P_{in}}{P_{out}}\right) + 2\ln\left(\frac{T_{in}}{T_{out}}\right)\right] \tag{8}$$

where L is the length, T_{av} is the average temperature of the gas between the inlet and outlet and Ω represents the cross-sectional area of the microchannel. Hydraulic diameter was defined as

$$D_h = \frac{2\Omega}{w+h} \tag{9}$$

where w and h are width and height of the microchannel, respectively. Finally, Reynolds number at the microchannel inlet was defined as $Re = \frac{m_\mu D_h}{\mu \Omega}$, where μ represents the dynamic viscosity of the nitrogen gas at the inelt. Two separate theoretical correlations were used to compare the experimental friction factor for each of the laminar and turbulent flow regimes. For a fully developed flow in a rectangular channel, the laminar friction factor can be calculated using the Shah & London (S&L) correlation as [31,32]

$$F_{\{SL\}} = \frac{96\left(1 - 1.3553\,\beta + 1.9467\beta^2 - 1.7012\beta^3 + 0.9564\beta^4\right)}{Re} \tag{10}$$

where aspect ratio β is defined as the ratio of the depth over the width of the microchannel. The experimental friction factor was compared to the Blasius law in the turbulent regime, which is given as follows:

$$F_{Turb} = 0.3164\,Re^{-0.25} \tag{11}$$

CFD simulations were also performed on microchannels with perturbators for the specific experimental mass flow rates. The temperature drop was negligible and not taken into account for CFD analysis. A single side of the counter-flow channels was taken for the pressure-loss studies, and 3D-Ideal gas equations were utilized using the K-ω turbulence model for CFD analysis. The microchannel length (L) with the S-shaped perturbator was 74 mm. An additional length of 9.696 mm was added to the inlet side, and 16.575 mm was added to the outlet part. This was implemented to fit the existing testbench. In conclusion, the microchannels were 101 mm in length, with 74 mm having S-shaped perturbators.

A comparison between the experimental and numerical pressure loss is shown in Figure 9a. The CFD and experimental results coincided with reasonable accuracy (<4%). The primary vertical axis denotes the pressure losses and the secondary vertical axis shows the deviation of CFD from experiments in %. The maximum deviation was larger at lower mass flow rates where the turbulent transition occurred. The RANS turbulent transition model was not reasonably accurate to predict turbulent transition. At higher mass flow rates, the CFD pressure losses were more reliable (<2%) when the flow was fully turbulent. The experimental friction factor evaluated using Equation (8) is shown in Figure 9b. The perturbators cause an increased surface roughness in the microchannels. Thus, as expected, the evaluated friction factor stayed higher than the S&L correlation in the early laminar regime. The slope of the experimental friction factor curve started deviating from the slope of laminar friction factor (S&L) and became parallel to the Blasius correlation at a considerably lower Re when compared to a smooth microchannel. The deviation in the current case began at an Re ~300, as shown in Figure 9b. This implies that the anticipated turbulent transition occurred at a lower Re, and is in agreement with several studies conducted by Shah and Wanniarachchi [11] and Heggs [12] on corrugated plate heat exchanger microchannel geometries. For Shah, the critical Reynolds number was less than 200, while Heggs [12] suggests that flow is never laminar after Re = 150. The length scale in this paper refers to the surface roughness and perturbator size in microchannels. If the length scale was too small, the turbulent transition occurred at the conventional Reynolds number. As the length scale increased, the critical Reynolds number reduced to a lower Reynolds number.

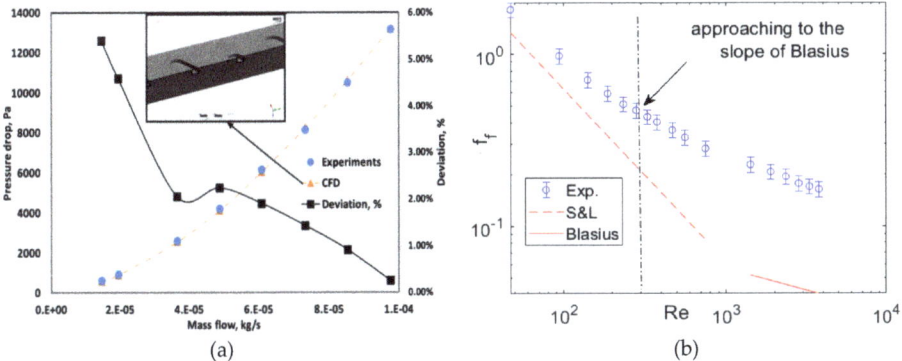

Figure 9. Experimental and CFD pressure-loss comparison and deviations (%) of the CFD results from the experimental pressure losses (dashed lines are for the numerical results and symbols are for the experimental readings) (**a**); experimental friction factor (**b**).

4.2. Collector

A collector with a microchannel configuration was fabricated using an Ultimaker 3+ extended version 3D printer based on fused deposition modeling (FDM). An open source software known as "Cura" was used to generate the 3D printer slicing application. The 3D printer transformed the design into several cross-sections then provided successive layers, until a three-dimensional object was formed. The plastic filaments were fed into the printer and the melted plastic extruded through the nozzle head and was directed on the building plate, where it created an object layer by layer. The biodegradable polylactic acid (PLA) plastic material was selected because PLA is having additives such as color pigments, plasticizers, nucleating agents and so on to optimize the properties for the specific application. The 3D printer had two nozzles: a primary nozzle to print the body and a secondary nozzle to print supporting structures for better surface finish and quality printing. Calibration was carried out to level the building plate. The building plate is the plate in which the microchannel is extruded. The building plate was heated to 40 °C. In addition, the nozzle was heated to 210 °C. In order to avoid the adhesion issues of the first layer, normal glue was provided on the building plate. The quality of the first layer was paramount, and it was guaranteed that the first layer had nicely adhered onto the glass plate with flat lines of filament. PLA prints at moderate temperatures, specifically 210 °C, depend on the selected nozzle size and print profile. Profiles for 0.25-mm nozzles use a slightly lower temperature. A temperature of 60 °C was maintained on the building plates. An ultrasonic cleaner was used for postprocessing to remove the supporting materials. However, it was difficult to remove the supporting structures from the microchannels. Apparently, the primary nozzle AA was only used for the collector fabrication. In order to investigate the surface finish and avoid overhang structures, two layers of microchannels were fabricated separately and glued. Figure 10a shows the experimental test design without collector covering. The steps in the collectors were introduced to decrease the flow maldistribution.

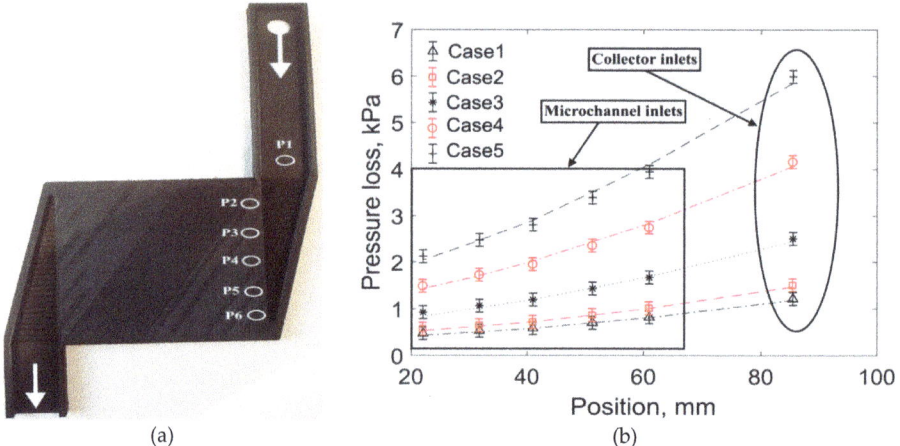

Figure 10. Experimental test design for collector flow maldistribution (**a**); comparison plot between experimental and CFD pressure losses (dashed lines are for the numerical results and symbols are for the experimental readings) (**b**).

An experimental test apparatus of secondary collectors with microchannels is depicted in Figure 10a. The inlet of the secondary collectors was extruded to have a uniform flow at the collector inlets. The extruded height of the collector inlet was selected based on a CFD optimization study. Six pressure taps (P1–P6) were installed for pressure readings (see Figure 10a). Five pressure taps were installed near the microchannel inlets (P2–P6), and the remaining one (P1) was placed at the collector inlet covering. Experimental testing was carried out for five different cases. The inlet mass flow rates are shown in Table 2. The size of the pressure tap hole was 60 microns, and it was ensured that the pressure taps were placed precisely at the center of the duct.

Table 2. Collector inlet mass flow rates for different cases.

Cases	Case 1	Case 2	Case 3	Case 4	Case 5
Mass flow rate, kg/s	3.01×10^{-4}	3.36×10^{-4}	4.32×10^{-4}	5.54×10^{-4}	6.66×10^{-4}

It is relatively easy to measure the mass flow rate of selected microchannels in a heat sink or heat exchanger if liquids are used as the working fluid [28]. Moreover, optical techniques such as micro particle image velocimetry can also be employed to quantify the maldistribution of liquid flows. For a multichannel heat exchanger with gas as the working medium, however, it is challenging to measure the mass flow rate of each microchannel. Therefore, an indirect method was used in the current work where pressure loss was measured across five microchannels amongst 33 parallel microchannels. An unequal pressure profile in the trapezoidal manifold suggests an opposite profile of maldistribution, as the mass flow through each microchannel was directly linked with the associated pressure loss. Experiments were therefore carried out to investigate the collector pressure losses and compare them with CFD. The results of the comparison between CFD and the microchannel measurements are shown in Figure 10b. Microchannel inlets (P2) near the collector inlets had more pressure losses. On the contrary, the microchannel inlets (P6) that were far away from the collector inlets had lower pressure losses. The pressure-loss decreasing pattern (from the collector inlet (P1) to the last microchannel (P6)) was nearly quadratic. In addition, there was a good agreement between CFD and the experiments, and the relative error was less than 4%. This validates the employed turbulence model in the collector studies presented earlier, and therefore the maldistribution pattern predicted by the CFD model can be considered to be representative of reality.

5. Conclusions

The influences of wire-net and S-shape perturbators on microchannel and collector performances were evaluated separately based on a Reduced Order Model (ROM). Microchannels with a wire-net were considered for comparative studies. Furthermore, the microchannel and collector performances were experimentally analyzed based on pressure-loss studies. The wire-net microchannels showed a better performance in terms of thermal efficiency, while S-shape fins provided a better performance in terms of pressure loss. Collectors with wire-net microchannels provided a more uniform flow distribution compared to the S-shape microchannels. Collector flow maldistribution decreased with higher microchannel resistance (i.e., smaller microchannels).

The recirculation zones in the collectors influenced the flow maldistribution. The strength of the recirculation zone near the collector beds was suppressed at higher mass flow rates. From the primary collector analysis, it was found that there was an optimum height where the pressure losses were saturated and the collector maldistribution decreased. The choice of a turbulent model was verified by comparing the CFD results with experimental tests performed for a single microchannel with S-shape perturbators and a micro heat exchanger with triangular collectors. Experimental testing of adiabatic flow showed that as a result of perturbators the onset of turbulent transition shifts to lower Reynolds number values compared to free channels. Apart from that, the microchannel and collector simulation results coincided with the experimental results with good accuracy (<4%).

Author Contributions: Simulations and Analysis, J.J.; Supervision, M.D., J.J.B., R.N. and G.L.M.; Funding acquisition, M.D., J.J.B., G.L.M. and G.J.K.; Experimental Investigation, D.R. and J.J.; Validation, J.J. and D.R.; Writing—original draft, J.J.; Review: J.J.B., J.G.K., M.D. and D.R. All authors have read and agreed to the published version of the manuscript.

Funding: This ITN Research Project MIGRATE (www.migrate2015.eu) is supported by European Community H2020 Framework under the Grant Agreement No. 643095. The research of MITIS has been funded by grants Nanocogen+ and Nanocogen+2 from DGO4 and DGO6 directorates of Wallonia.

Acknowledgments: Microfabrication of microchannels with S-shaped perturbators from Institute of Microstructure Technology is acknowledged. Technical support of Franceso Vai for fabrication of the MC test section assembly is acknowledged.

Conflicts of Interest: The authors declare no conflict of interest.

Abbreviations

A	cross-sectional area, m^2
BCs	boundary conditions
CFD	computational fluid dynamics
CHP	combined heat and power
CHT	conjugate heat transfer
$Cp_{c/h}$	heat capacity at constant pressure for cold/hot inlet microchannels, J/(K.kg)
C_c	cold fluid capacity rate, W/(K)
C_h	hot fluid capacity rate, W/(K)
D_h	hydraulic diameter, mm

L	microchannel length, mm
m_μ, $m_{\mu c}$, $m_{\mu h}$	microchannel mass flow rate for cold/hot inlet channels, $\frac{kg}{s}$
m_{co}	collector inlet mass flow rate for cold/hot inlet side, $\frac{kg}{s}$
$m_{\mu CHT}$	microchannel mass flow rate calculated from CHT model inlet, $\frac{kg}{s}$
w	width of the microchannel, mm
h	height of the microchannel, mm
n_{pl}	number of plates
PM	porous medium
P, P_{in}, P_{out}, P_{ab}	inlet/outlet total pressure and absolute pressure, Pa
ROM	reduced order model
R	universal gas constant, $\frac{J}{mol\,k}$
Re	Reynolds number
S_1	Sutherland constant
T	temperature, K
$T_{in,out}^{h,c}$	inlet and outlet total temperatures for hot \- cold microchannel inlets and outlets, K
T_{av}	is the average temperature of the gas between the inlet and outlet, K
ΔT	temperature drop across the microchannel length, K
ΔP	pressure losses across the length, Pa
Q	thermal source term, $\frac{W}{m^3}$
U, U_{in}	stream wise mean velocity, inlet velocity, $\frac{m}{s}$
$1/\alpha$	viscous coefficient, m^{-2}
β	inertial coefficients, m^{-1}
$\beta 1$	ratio of the depth over the width of the microchannel
Υ	microchannel size reduction coefficient
ρ	density, kg/m^3
μ	dynamic viscosity of the Nitrogen gas, kg/(m·s)
ε	thermal efficiency, %
ε_m	turbulent viscosity, $m^2 \cdot s^{-1}$

References

1. Min, J.K.; Jeong, J.H.; Ha, M.Y.; Kim, K.S. High temperature heat exchanger studies for applications to gas turbines. *Heat Mass Transf.* **2009**, *46*, 175–186. [CrossRef]
2. Xiao, G.; Yang, T.; Liu, H.; Ni, N.; Ferrari, M.L.; Li, M.; Luo, Z.; Cen, K.; Ni, M. Recuperators for micro gas turbines: A review. *Appl. Energy* **2017**, *197*, 83–99. [CrossRef]
3. Shah, R.K. Advances in Science and Technology of Compact Heat Exchangers. *Heat Transf. Eng.* **2006**, *27*, 3–22. [CrossRef]
4. Kandlikar, S.G.; Garimella, S.; Li, D.; Colin, S.; Michael, R.K. *Heat Transfer and Fluid Flow in Minichannels and Microchannels*; Elsevier Science: Amsterdam, The Netherlands, 2014; Volume 2, pp. 11–102.
5. Siva, V.M.; Pattamatta, A.; Das, S.K. A Numerical Study of Flow and Temperature Maldistribution in a Parallel Microchannel System for Heat Removal in Microelectronic Devices. *J. Therm. Sci. Eng. Appl.* **2013**, *5*, 041008. [CrossRef]
6. Ayyaz, S.; Amit, S.; Anugrah, K.S.; Sandip, J.; Bhaskar, M. Design of a collector shape for uniform flow distribution in microchannels. *J. Micromech. Microeng.* **2017**, *27*, 075026.
7. Tao, W.; He, Y.; Wang, Q.; Qu, Z.; Song, F. A unified analysis on enhancing single phase convective heat transfer with field synergy principle. *Int. J. Heat Mass Transf.* **2002**, *45*, 4871–4879. [CrossRef]
8. Fiebig, M. Vortex generators for compact heat transfer exchangers. *J. Enhanced Heat Transfer* **1995**, *2*, 43–61. [CrossRef]
9. Valencia, A. Turbulent flow and heat transfer in a channel with a square bar detached from the wall. *Numer. Heat Transfer, Part A: Appl.* **2000**, *37*, 289–306. [CrossRef]
10. Caccavale, P.; De Bonis, M.V.; Ruocco, G. Conjugate heat and mass transfer in drying: A modeling review. *J. Food Eng.* **2016**, *176*, 28–35. [CrossRef]
11. Shah, R.; Heikal, M.; Thonon, B.; Tochon, P. Progress in the numerical analysis of compact heat exchanger surfaces. *Adv. Heat Transfer* **2001**, *34*, 363.

12. Heggs, P.; Sandham, P.; Hallam, R.; Walton, C. Local Transfer Coefficients in Corrugated Plate Heat Exchanger Channels. *Chem. Eng. Res. Des.* **1997**, *75*, 641–645. [CrossRef]
13. Focke, W.; Zachariades, J.; Olivier, I. The effect of the corrugation inclination angle on the thermohydraulic performance of plate heat exchangers. *Int. J. Heat Mass Transf.* **1985**, *28*, 1469–1479. [CrossRef]
14. Liu, F.-B.; Tsai, Y.-C. An experimental and numerical investigation of fluid flow in a cross-corrugated channel. *Heat Mass Transf.* **2010**, *46*, 585–593. [CrossRef]
15. Alfa Laval, Heat Exchanger Fouling. Available online: https://www.alfalaval.com/globalassets/documents/industries/pulp-and-paper/ppi00424en.pdf (accessed on 19 March 2020).
16. Yang, Y.; Gerken, I.; Brandner, J.J.; Morini, G.L. Design and Experimental Investigation of a Gas-to-Gas Counter-Flow Micro Heat Exchanger. *Exp. Heat Transf.* **2014**, *27*, 340–359. [CrossRef]
17. Yang, Y.; Brandner, J.; Morini, G.L. Hydraulic and thermal design of a gas microchannel heat exchanger. *J. Physics: Conf. Ser.* **2012**, *362*, 012023. [CrossRef]
18. Joseph, J.; Delanaye, M.; Nacereddine, R.; Giraldo, A.; Rouabah, M.; Korvink, J.G.; Brandner, J.J. Numerical Study of Perturbators Influence on Heat Transfer and Investigation of Collector Performance for a Micro-Combined Heat and Power System Application. *Heat Transf. Eng.* **2020**, 1–23. [CrossRef]
19. Joseph, J.; Delanaye, M.; Nacereddine, R.; Giraldo, A.; Rouabah, M.; Korvink, J.G.; Brandner, J.J. Numerical and experimental investigation of a wire-net compact heat exchanger performance for high-temperature applications. *Appl. Therm. Eng.* **2019**, *154*, 208–216. [CrossRef]
20. Joseph, J.; Nacereddine, R.; Delanaye, M.; Korvink, J.G.; Brandner, J.J. Advanced Numerical Methodology to Analyze High-Temperature Wire-Net Compact Heat Exchangers for a Micro-Combined Heat and Power System Application. *Heat Transf. Eng.* **2019**, 1–13. [CrossRef]
21. Delanaye, M.; Giraldo, A.; Nacereddine, R.; Rouabah, M.; Fortunato, M.V.; Parente, A. Development of a recuperated flameless combustor for an inverted bryton cycle microturbine used in residential micro-CHP. Proceedings of ASME Turbo Expo 2017: Turbomachinery Technical Conference and Exposition, Charlotte, NC, USA, 26–30 June 2017.
22. Manceau, R. *Industrial Codes for CFD, International Masters in Turbulence*, 1st ed.; Applied Mathematics Departement, Inria-Cagire Group, CNRS–University of Pau: Pau, France, 2017; Available online: http://remimanceau.gforge.inria.fr/Publis/PDF/IndustrialCodesForCFD.pdf (accessed on 23 May 2019).
23. Morini, G.L.; Lorenzini, M.; Colin, S.; Geoffroy, S. Experimental Analysis of Pressure Drop and Laminar to Turbulent Transition for Gas Flows in Smooth Microtubes. *Heat Transf. Eng.* **2007**, *28*, 670–679. [CrossRef]
24. Rehman, D.; Joseph, J.; Morini, G.L.; Delanaye, M.; Brandner, J. A Hybrid Numerical Methodology Based on CFD and Porous Medium for Thermal Performance Evaluation of Gas to Gas Micro Heat Exchanger. *Micromachines* **2020**, *11*, 218. [CrossRef]
25. Yang, Y. Experimental and Numerical Analysis of Gas Forced Convection through Microtubes and Micro Heat Exchangers. Ph.D. Thesis, Ingegneria Energetica, Nucleare e del Controllo Ambientale, Unibo, Italy, 2013.
26. Gersten, K.; Schlichting, H. *Boundary-Layer Theory*, 9th ed.; Springer: Berlin/Heidelberg, Germany, 2017.
27. Grid turbulence. Available online: http://www.lfpn.ds.mpg.de/turbulence/generation.html (accessed on 19 March 2020).
28. Anbumeenakshi, C.; Thansekhar, M. Experimental investigation of header shape and inlet configuration on flow maldistribution in microchannel. *Exp. Therm. Fluid Sci.* **2016**, *75*, 156–161. [CrossRef]
29. Teng, J.-T.; Chu, J.-C.; Liu, M.-S.; Wang, C.-C.; Greif, R. Investigation of the Flow Mal-Distribution in Microchannels. In *Advances in Bioengineering*; ASME International: New York, NY, USA, 2003; Volume 2, pp. 465–469.
30. Renault, C.; Colin, S.; Orieux, S.; Cognet, P.; Tzedakis, T. Optimal design of multi-channel microreactor for uniform residence time distribution. *Microsyst. Technol.* **2011**, *18*, 209–223. [CrossRef]

31. Rehman, D.; Morini, G.L.; Hong, C. A Comparison of Data Reduction Methods for Average Friction Factor Calculation of Adiabatic Gas Flows in Microchannels. *Micromachines* **2019**, *10*, 171. [CrossRef]
32. Sahar, A.M.; Wissink, J.; Mahmoud, M.; Karayiannis, T.; Ishak, M.S.A.; Ishak, I.D.M.S.A. Effect of hydraulic diameter and aspect ratio on single phase flow and heat transfer in a rectangular microchannel. *Appl. Therm. Eng.* **2017**, *115*, 793–814. [CrossRef]

© 2020 by the authors. Licensee MDPI, Basel, Switzerland. This article is an open access article distributed under the terms and conditions of the Creative Commons Attribution (CC BY) license (http://creativecommons.org/licenses/by/4.0/).

Article

Numerical Thermal Analysis and 2-D CFD Evaluation Model for An Ideal Cryogenic Regenerator

Natheer Almtireen [†], Jürgen J. Brandner and Jan G. Korvink *

Institute of Microstructure Technology (IMT), Karlsruhe Institute of Technology (KIT), 76344 Eggenstein-Leopoldshafen, Germany; Natheer.Almtireen@kit.edu (N.A.); Juergen.Brandner@kit.edu (J.J.B.)
* Correspondence: Jan.Korvink@kit.edu; Tel.:+49-721-608-22740
† Scholarship holder to complete doctorate degree; granted by the German Jordanian University (GJU).

Received: 21 February 2020; Accepted: 27 March 2020; Published: 30 March 2020

Abstract: Regenerative cryocoolers such as Stirling, Gifford–McMahon, and pulse tube cryocoolers possess great merits such as small size, low cost, high reliability, and good cooling capacity. These merits led them to meet many IR and superconducting based application requirements. The regenerator is a vital element in these closed-cycle cryocoolers, but the overall performance depends strongly on the effectiveness of the regenerator. This paper presents a one-dimensional numerical analysis for the idealized thermal equations of the matrix and the working gas inside the regenerator. The algorithm predicts the temperature profiles for the gas during the heating and cooling periods, along with the matrix nodal temperatures. It examines the effect of the regenerator's length and diameter, the matrix's geometric parameters, the number of heat transfer units, and the volumetric flow rate, on the performance of an ideal regenerator. This paper proposes a 2D axisymmetric CFD model to evaluate the ideal regenerator model and to validate its findings.

Keywords: cryogenics; MATLAB®; numerical thermal analysis; cryocooler; regenerator; optimization; ANSYS Fluent

1. Introduction

The growth in a large number of low-temperature applications led to huge developments in cryogenics and particularly cryocoolers over the past few decades. Top applications among these are IR imaging, i.e., micro-cryocoolers for IR imaging systems [1], superconductivity based systems, i.e., cooling of high temperature superconducting (HTS) motors [2], aerospace and military applications, cryosurgery, i.e., cryoablation of locations in the heart to treat heart arrhythmia [3], and many others. Some merits of these cryocoolers are comprised of durability, ruggedness, compactness, fast response, low vibration, and for the case of pulse tube cryocoolers (PTC), the absence of a mechanical displacer and relatively rapid cooling. Generally, cryocoolers utilize the oscillatory compression and expansion of a gas, commonly helium, within a closed volume to cool down an object. In general, any useful attempt to bring cryocoolers, especially pulse tube cryocoolers, down to the miniaturized scale requires a thorough study of each component's working principle, since the cryocooler performance is highly dependent on the efficiency of each component involved in the thermodynamic cycle.

The regenerator is a vital component in any closed-cycle regenerative cryocooler, including the Stirling, Gifford–McMahon (G-M), and pulse tube machines. In pulse tube cryocoolers, the regenerator acts as a thermal storage component that transmits the pressure signal from the compressor to the pulse tube and other cryocooler components. It comprises a metallic porous medium that is subject to the periodic flow of the gas; where it is conventionally constructed of a thin hollow tube filled with mesh screens and/or metallic spheres, etc. The working principle involves thermal energy exchange between the working gas and the matrix material. To maintain considerable refrigeration, this matrix takes away the heat from the incoming gas during the heating phase and delivers it back for the returning gas during the cooling phase. The frequency of the gas flow can reach 100 Hz for a conventional PTC, but this should be significantly increased to serve any future miniaturized PTC design.

In 1816, Stirling introduced the regenerative heat exchanger in a hot air engine [4]. In 1927, Nusselt presented a mathematical analysis for a special case regenerator with infinite matrix heat capacity [4]. Iliffe used the ideal analyses of several German authors to perform a first-order graphical numerical analysis that accounted for time variations of the matrix temperature as early as 1948 [5]. One of the major advances happened in 1960, when Gifford and McMahon developed a small regenerative cryogenic refrigerator to cool infrared detectors and maser amplifiers, which moved cryogenics out of the laboratory to industry [4]. Kuriyama et al. [6] used rare-earth regenerative materials to achieve ultra-low temperatures with a G-M multi-stage cryocooler.

Analytical and numerical analyses of critical components are required for the development of more efficient pulse tube cryocoolers, to replace potentially conventional concepts in existing or emerging application fields [7–10]. In particular, these numerical analyses and thorough experimental works are very significant in providing better interpretation of the thermal interaction between the working gas and the matrix material in the regenerator. Radebaugh [11] introduced an optimization procedure for designing regenerators. The REGEN [12] series software developed at the National Institute of Standards and Technologies (NIST) was used by many researchers to study the regenerator performance for different matrix materials. The model in REGEN3.2 assumes sinusoidal mass flow at both ends of the regenerator, where the incoming gas temperature for both ends and the gas pressure are initially stated; also, the initial temperature profile is generally considered a linear profile. Further, the model sets the amplitude, frequency, and phase of the mass flow at both ends as input parameters. However, the equation that is used to describe the flow is not the conservation equation, but rather a predictive equation for pressure. Additionally, The gas velocity for all mesh cells is calculated simultaneously by solving a non-linear system; the remaining variables at each mesh cell are advanced in time explicitly. Hence, the model numerical approximation is considered semi-implicit [13]. On the other hand, REGEN3.3 is referred to as a fully-implicit model; since the numerical approximation involves discretization of the conservation of mass, momentum, and energy differential equations, whereas the non-linear system of equations for temperature, pressure, and mass flow at all mesh points are solved simultaneously by Newton iteration [13]. Other available software packages including SAGE® and DELTAE®, which solve the conservation equations and hence describe the regenerator performance, are also often used. These numerical analyses have a range of complexity levels that depend on the assumptions employed and the degree of freedom used.

For all complexity levels, the design parameters of the regenerator, such as its diameter-to-length ratio, physical dimensions, pore structure, and matrix material parameters, impact the overall performance for any regenerative cryocooler. Normally, the design and/or optimization procedure of these parameters have been empirical, based on relatively rough lumped parameter considerations or the results of dimensional analysis. However, more complex numerical models should account for pressure drop, void volume, conduction losses, pressure variations, time-varying mass flow rates, time-varying inlet gas temperature, and temperature-dependent thermal properties. This paper presents a numerical approximation for the thermal interaction between the gas and the matrix material by only solving the energy conservation equations for both the matrix and the gas; the assumptions involve uniform one-dimensional longitudinal flow, with infinite radial thermal conduction and zero longitudinal thermal conduction in the matrix. It numerically solves the idealized general thermal equations for the working gas and the matrix based on finite-difference techniques and thus predicts the output temperature profiles for the working gas and the matrix material. Moreover, it studies the efficiency of the regenerator versus several design parameters of the regenerator. Finally, to validate the findings of the ideal regenerator model, the paper presents a 2D axisymmetric CFD model that solves the mass, momentum, and energy equations for the porous regenerator zone and provides more accurate results to compare it with the ideal regenerator model and examine its soundness.

2. Regenerator System

The regenerator is nearly analogous to a lossy spring that alternatively accepts energy and releases it when the excitation is removed. Another analogous model is a thermal flywheel accepting energy on the down-stroke and giving it back on the upstroke [14]. In the regenerator segment, the matrix is contained within a conventional steel tube housing in order to withstand the oscillatory high-pressure stream. The matrix should have an excellent thermal storage capacity with large heat transfer area, along with imposing a minimum pressure drop on the flow. Unfortunately, the last two characteristics are rather contradictory, since maximizing heat transfer area requires reducing the free flow area, which results in decreasing the gas volume and causing larger pressure drop. The advantages of regenerative heat exchangers can be listed as follows [14]:

- A large area for heat transfer can be obtained using inexpensive finely stacked material.
- The construction is relatively straightforward, and considerable savings can be made for the same heat exchanger duty.
- The regenerator tends to be self-cleaning due to the nature of the periodic flow reversals, and this is advantageous if contaminated gas is being processed.
- Well-designed regenerators can be achieved if proper design optimization procedures are in place.

However, the major disadvantage for static regenerative exchangers is the mixing of the hot and cold streams and is inescapable because of the carryover in flow switching [14]. This could disrupt the heat exchanging process and therefore deteriorate the cooling process; thus, great care should be taken when designing or constructing the regenerator element for optimum operation.

In the regenerator, the temperatures of the gas and the pores vary with location and time. However, after several cycles of heating and cooling, a periodic steady state is achieved, as a result the nodal temperature at any location in the regenerator is equal to that temperature before/after one complete cycle at the same location. The expected temperature distributions for the gas and the matrix over one cycle are shown in Figure 1. It is worth mentioning that the cryocooler system itself is a closed system, in which no mass transfer out/to the system is allowed, and only energy transfer is permitted. Therefore, any indication of gas inflow/outflow refers to flow inside the system from one component to/from the regenerator.

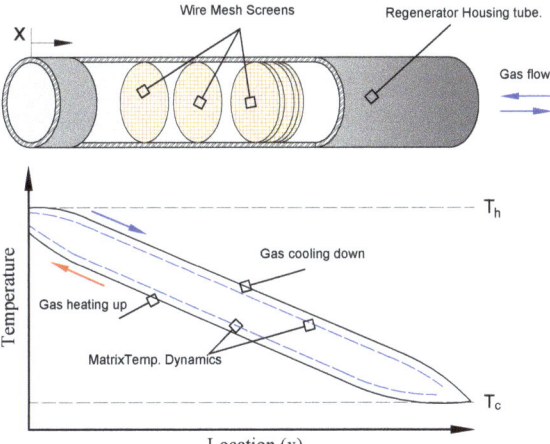

Figure 1. Expected matrix and gas temperature dynamics, where T_h and T_c are the respective hot and cold temperature levels.

2.1. Ideal Regenerator Model and Numerical Solution

The ideal regenerator can be interpreted as a thermodynamic object that receives the gas during the heating phase with low temperature T_c and releases it with high temperature T_h. After the flow direction is reversed, the gas enters with temperature T_h and leaves with T_c. In practice, the ideal regenerator case is impossible to achieve, where maintaining constant inlet/outlet temperature would need infinitely slow operation, or an infinite heat transfer coefficient, and/or a great heat transfer area. Furthermore, the heat capacity of the fluid and the matrix should be respectively zero and infinity. Moreover, the absence of a pressure drop would require a frictionless flow, while the absence of void volume would prevent the provision of flow passages through the matrix for the fluid to traverse [14].

The mathematical model used for this publication is based on simplified thermal energy conservation for the working gas and the matrix, as suggested by Ackermann [4]. The one-dimensional model is built on various assumptions: first, the mass flow and the working gas pressure through the regenerator are constant, and their magnitudes are equal during the heating and cooling periods, with no gas flow-mixing during these phases. In other words, the hot gas enters the regenerator at a constant temperature with constant and uniform velocity over the cross-sectional area; it stops when the flow is reversed, and the cold gas enters the regenerator with a constant temperature and flow rate. Second, the boundary temperature conditions for the working gas are constant for heating and cooling periods. Third, the fluid stored thermal energy is zero, with zero longitudinal matrix thermal conductivity. As a result, the matrix and fluid thermal equations are:

$$\text{Matrix balance equation: } h\delta A_s \left(T_g - T_m\right) = (\rho c_p \delta V)_m \frac{\partial T_m}{\partial t}, \tag{1}$$

$$\text{Fluid balance equation: } h\delta A_s \left(T_g - T_m\right) = -(\rho c_p \delta V)_g u_x \frac{\partial T_g}{\partial x}. \tag{2}$$

Here, h, A_s, T_m, T_g, ρ, u_x, δV, and c_p are, respectively, the convection heat transfer coefficient, matrix heat transfer area, matrix temperature, gas temperature, the density of the gas, the velocity of the gas, the control volume, and constant pressure specific heat capacity. The left-hand sides for both equations represent the convection heat transfer between the matrix material and the working gas, while the right-hand sides in Equations (1) and (2) represent the varying built-up heat in the matrix and the working gas. Equations (1) and (2) are then expressed in finite-element difference form as proposed by Ackermann [4], where the regenerator is divided into spatial elements and the heating and cooling periods are divided into sufficiently small time steps. Figure 2 shows the 1D computational mesh for the regenerator during heating and cooling periods.

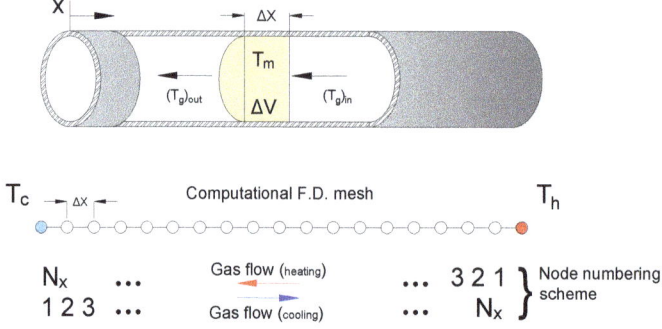

Figure 2. Numerical schematic for the regenerator during heating and cooling periods, where N_x and N_t are the total number of spatial and time elements and T_h and T_c are the hot and cold temperature levels.

The model is then converted to two linear algebraic equations, which are used to compute the nodal temperatures for the matrix and the gas, hence predicting the regenerator performance and finally yielding:

$$\text{Matrix time marching scheme: } (T_g)_{i+1}^{j} = (T_g)_i^j - K_1((T_g)_i^j + (T_m)_i^j), \tag{3}$$

$$\text{Fluid time marching scheme: } (T_m)_i^{j+1} = (T_g)_i^j + K_2((T_g)_i^j + (T_m)_i^j). \tag{4}$$

Here, i and j are the number of spatial and time nodes, and K_1 and K_2 are constants [4]:

$$K_1 = \left(\frac{\left(\frac{hA_s}{\dot{m}c_p}\right)_f \Delta x}{1 + \left(\frac{hA_s}{(Mc_p)_m}\right)\Delta t\right)_{\Delta x} + \frac{1}{2}\left(\frac{hA_s}{\dot{m}c_p}\right)_f \Delta x} \right)$$

$$K_2 = \left(\frac{\left(\frac{hA_s}{(Mc_p)_m}\right)\Delta t}{1 + \left(\frac{hA_s}{(Mc_p)_m}\right)\Delta t\right)_{\Delta x} + \frac{1}{2}\left(\frac{hA_s}{\dot{m}c_p}\right)_f \Delta x} \right)$$

Here, $\dot{m}$, M, Δt, and Δx are the gas mass flow rate, the mass of the matrix material, and the time and the space differential elements, respectively. The equations are solved in a sequential fashion for each spatial and time node during both the heating and cooling periods. The numerical scheme assumes a linear distribution of temperature between T_h and T_c across the matrix. Furthermore, the fluid is assumed to enter the regenerator at T_h for the heating period and T_c for the cooling period. The MATLAB® code calculates the nodal temperatures for both the matrix and gas during the heating period. The output spatial nodal temperatures for the matrix at the end of the heating period are then used as the input spatial nodal temperatures for the matrix at the start of the cooling period, and this can be expressed mathematically as:

$$(T_m)_{1:N_x}^1 \Big/_{Cooling} = (T_m)_{N_x:1}^{N_t} \Big/_{Heating}$$

For the algorithm to converge to a solution, an increased number of time steps is needed; generally, the number of spatial nodes should be larger than the total number of heat transfer units (NTU). The relation between the regenerator inefficiency (Ie) and the spatial and time steps is discussed in Sections 2.2 and 3.1.

The regenerator's effectiveness term measures the performance of the heat exchange process relative to ideal heat exchanging conditions. It is defined as the ratio of the actual heat exchanged between fluids and media to the ideal exchanged heat if no temperature difference exists between the two fluids at any position. The expression of the regenerator's effectiveness in the case of balanced inlet and outlet flows is [4]:

$$\epsilon = \frac{T_{h,in} - \overline{T}_{h,out}}{T_{h,in} - T_{c,in}} = \frac{\overline{T}_{c,out} - T_{c,in}}{T_{h,in} - T_{c,in}} \tag{5}$$

where $T_{h,in}$, $T_{c,in}$, and $T_{c,out}$ are the temperature of the inlet heating fluid, the average outlet cooling fluid, and the temperature of the inlet cooling fluid, respectively. Generally, the performance of a cryogenic regenerator is defined rather by the inefficiency (Ie), where [4]:

$$\text{Inefficiency}(Ie) = 1 - \epsilon = 1 - \frac{T_{h,in} - \overline{T}_{h,out}}{T_{h,in} - T_{c,in}} \tag{6}$$

2.2. Model Convergence

The model converges to a solution under two conditions. During the heating cycle, the final matrix temperature should not be higher than the outlet heating fluid temperature. During the cooling period, the final fluid temperature is not higher than the cooling outlet fluid temperature. This can be summarized mathematically as:

$$\left[(T_m)_{N_x}^{N_t}\right]_{Heating} \not> \left[(T_f)_{N_x}^{N_t}\right]_{Heating},$$

$$\left[(T_f)_{N_x}^{N_t}\right]_{Cooling} \not> \left[(T_m)_{N_x}^{N_t}\right]_{Cooling}.$$

From a mathematical point of view, the maximum time interval Δt, for each successive nodal calculation, is achieved when the output matrix and fluid temperatures are equal. If the outlet temperatures go beyond that, the crossing of the outlet temperatures causes a sign-reversal of the inlet temperature difference for the next calculation cycle and hence causes a divergence of the numerical solution. To prevent this crossover and thus divergence, the model must either set the time interval Δt to a minimal value or introduce a larger number of spatial nodes. Figure 3 illustrates the convergence criteria during the cooling period. The previous two conditions are applied to Equations (1) and (2) to deduce the reduced convergence formula: $K_1 + K_2 = 1$.

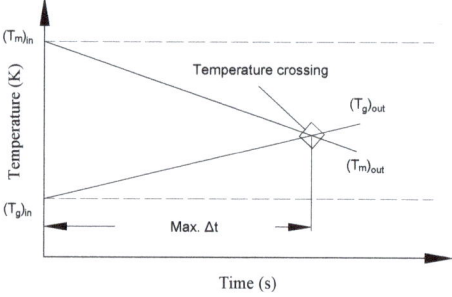

Figure 3. Convergence criteria for the numerical model during the cooling period.

The number of spatial nodes is also significant in defining the accuracy of the computed variables. As the number of spatial nodes increases, the final outlet gas temperature during the heating period approaches the inlet gas temperature during the cooling period, thus decreasing inefficiency (Ie), as suggested by Equation (5). Figure 4a shows the effect of increasing the number of spatial nodes on the accuracy of the inefficiency as a calculated variable, and Figure 4b shows the effect of the number of spatial nodes on the final outlet gas temperature. It shows that the proposed finite difference numerical model requires at least 4000 nodes to achieve tolerable accuracy if N_t is set to 200.

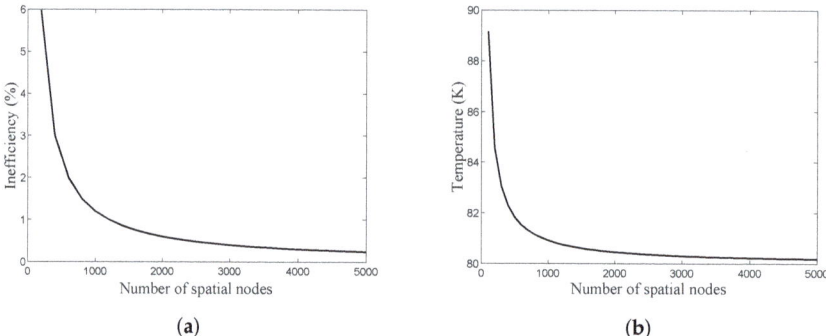

Figure 4. (a) Inefficiency (Ie) accuracy relative to the number of spatial nodes. (b) The accuracy in the final outlet gas temperature during the heating period both for the total number of heat transfer units (NTU$_{Total}$) of 63 and N_t of 200.

2.3. Model Application

The model is used to examine the effect of changing the regenerator dimensions (i.e., length and diameter) on the inefficiency (Ie). Second, it examines the implications of the matrix mesh screen parameters (i.e., wire diameter and mesh size) on the performance of the regenerator and its inefficiency. A particular case is highlighted, with the following parameters: a regenerator of 7.5 mm and 30 mm in diameter and length. The regenerator matrix mesh size is a 150 × 150 commercial phosphorous-bronze screen; the screen wire-diameter is 63 μm; and the screens have the same value as the regenerator diameter. Both the mesh size and wire diameter were used in the numerical model to calculate the hydraulic diameter for the screen, the porosity, the heat transfer coefficient, and the total heat transfer area A_s. Furthermore, if perfect stacking is assumed, the number of screens is set then to 238, and the cold and hot end temperatures are assumed to be 80 K and 300 K, respectively.

2.4. 2D Axisymmetric Regenerator Model

This section introduces a two-dimensional (2D) axisymmetric transient regenerator model to investigate the validity of the ideal regenerator model. The CFD model simulation was executed using ANSYS Fluent software since it is widely used for modeling and simulating fluid flows and heat transfer in various engineering problems. Although Fluent's momentum, continuity, and energy governing equations are inaccessible for modification, Fluent allows users to program their user-defined functions (UDF) and to connect them to the primary model to modify boundary conditions, alter domain properties, or introduce variable sources or signals.

The proposed system consisted of three zones: hot temperature zone, cold temperature zone, and porous regenerator zone. The model assumed that the hot temperature zone was contracting at a constant rate, pushing the hot gas inside the regenerator, while the cold temperature zone expanded at the same rate to maintain constant overall volume during the cooling period. After that, the hot zone started to expand, and the cold zone contracted to retain their original volumes as the cold gas entered the regenerator zone from the other side when the flow reversed direction during the heating period. The hot and cold temperature zones were designed to ensure that gas entered at T_h and T_c during the cooling and hot periods, respectively.

Figure 5 depicts the schematic diagram for the 2D model, where the dynamic mesh feature in Fluent, with C-language UDF, which defined the walls motion, were employed to enable the simultaneous motion of the left and right walls at constant speeds. The dynamic mesh feature allowed the user to create moving walls and deforming zones, which made any applications involving volume contraction and expansion modeled easily. In the dynamic mesh model, the layering option was employed to enable the addition or elimination of cells adjacent to a moving boundary based on the height of the layer adjacent to the moving surface in prismatic mesh zones [15]. The regenerator's dimensions were the same as the highlighted case in Section 2.3. The porous medium was modeled by the addition of a momentum source term to the standard fluid flow equations. This source term created a pressure drop that was proportional to the fluid velocity, and it was composed of two parts: a viscous loss and an inertial loss term that is represented as [15]:

$$(S_i) = \frac{\Delta p}{L} = -\left(\frac{\mu v_i}{\alpha} + C_2 \frac{\rho |v| v_i}{2}\right) \quad (7)$$

where L, α, C_2, μ, and v_i are the length of the regenerator, the permeability, the inertial resistance factor, the fluid viscosity, and the fluid velocity, respectively. The porous medium for this model was constructed of perfectly stacked #635 stainless steel screens with a wire diameter of 20.3 μm. The viscous and inertial resistance factors were applied to the porous zone, and their values were utilized from [16]. Landrum et al. estimated the hydrodynamic parameters of two fine-mesh porous materials (#325 and #635 mesh screens) by performing experiments, in which pressure variations across these mesh screens were measured under steady and oscillatory flows. Then, they simulated these experiments using a CFD model where the viscous and inertial resistances were iteratively adjusted until an agreement was reached between the experimental results and the simulated predictions [16].

The calculations reported here were performed using Intel® Core™ i7-5500U cores @ 2.4 GHz with four cores. The settings for the model were axisymmetric and laminar, and the working fluid was helium as an ideal gas. The numerical discretization schemes were second-order for the pressure, second-order upwind for both the momentum and energy, and with implicit first-order transient time formulation. The simulated flow time was 80 s, with a 2 ms time step size and 10 iterations per time step. The space element size in the dynamic mesh zones was set carefully to avoid negative volume calculation errors.

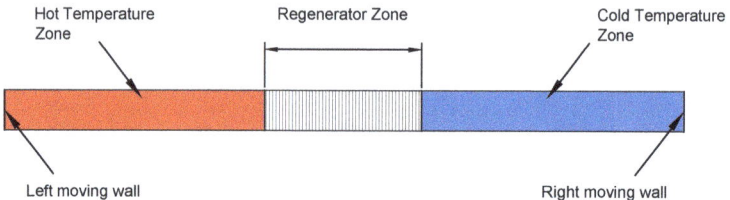

Figure 5. Two-dimensional regenerator model.

3. Results and Discussion

3.1. Ideal Regenerator

Figure 6a shows the gas and matrix temperature profiles during heating and cooling periods. It can be noticed that the temperature of the matrix material was increasing for all spatial elements during the heating period while the hot gas temperature dropped gradually as it crossed the regenerator, causing all the matrix elements to heat up. The cold gas was heated up as it passed the matrix, resulting in cooling the matrix. Figure 6b shows the algorithm after it converged and settled to a solution where the difference in temperature between the matrix and gas was minimal; this happened after increasing the number of time and/or space nodes, as described previously in Section 2.2.

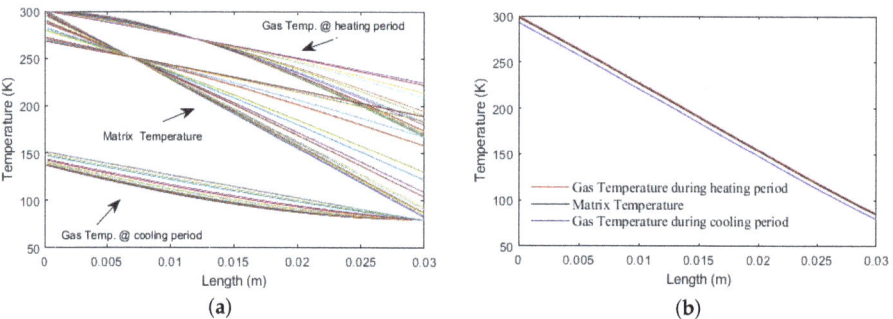

Figure 6. (a) Matrix and gas temperature variations during the heating and cooling period. (b) The final temperature profiles as they converge to a solution with increasing N_x and N_t.

The dimensional and parametric analyses are of central importance for any efficient regenerator design. Figure 7a shows the relation between the regenerator diameter and its inefficiency (Ie), for several lengths, while Figure 7b illustrates the relation between the length of the regenerator and the inefficiency (Ie) for several regenerator diameters. These were based on the ideal case interpretation of the thermal interaction between the working gas and the matrix material; hence, it did not consider conduction or viscous losses. Furthermore, it assumed that the thermal properties for both the matrix and the working gas were constant with temperature variation inside the tube.

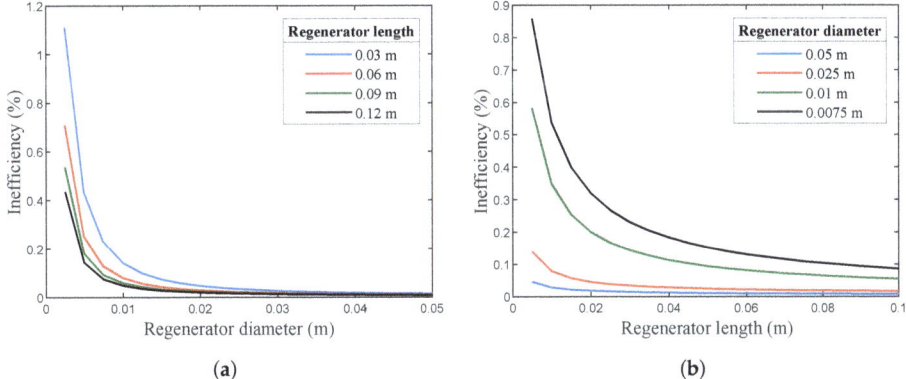

Figure 7. (a) The relation between the regenerator diameter and Ie. (b) The relation between the regenerator length and Ie (the matrix screen mesh size is 150 × 150 and has a wire diameter of 63 µm).

Although an abrupt change in the inefficiency was quite clear for diameters smaller than 0.01 m and lengths less than 0.03 m, the difference in the inefficiencies tended to be very small for diameter values above 0.025 m and lengths above 0.06 m. As a result, any attempt to miniaturize the regenerator element must take great care not to deteriorate its operation. The number of heat transfer units (NTU) is a non-dimensional parameter that expresses the regenerator's heat transfer capacity. It is directly related to the size of the regenerator, its flow, and thermodynamic considerations. In a sense, if the NTU was large, the effectiveness of the regenerator would be high. It could be noticed that the inefficiency dropped with the increase in regenerator length and diameter, i.e., NTU increased as well, since the mass of the regenerator and its total heat transfer area increased as well. It is worth noticing that, in non-ideal regenerators, the axial conduction and viscous losses increases with larger lengths, and as a result, the inefficiency (Ie) also decrease.

Mesh screens can be found in a variety of sizes (i.e., 20 × 20 to 500 × 500). They are commonly used in many heat exchanging setups in research and industrial applications, which include air vents, cryogenics equipment, cryocoolers, coldheads, and heat pipes. The notation 20 × 20 refers to 20 wires/openings per inch.

Figure 8a exhibits the relation of Ie with matrix mesh size. It was evident that as the screen mesh size increased, i.e., the number of openings per inch increased, both the porosity and the hydraulic diameter decreased; and in contrast, the thermal penetration depth decreased, causing the total heat transfer area to increase, hence resulting in decreasing the regenerator's inefficiency (Ie).

Figure 8b depicts the relation between the inefficiency (Ie) with the mesh screen wire diameter for several mesh sizes. It can be shown that the increase in the wire diameter decreased the inefficiency (Ie), since it decreased the porosity of the mesh screen, which was in turn inversely proportional to the mass of regenerator, hence leading to increasing the total heat transfer area. It is evident from Figure 8 that the regenerator's inefficiency dropped with the increase in the wire diameter of the regenerator for the same mesh size, and this was also valid for different mesh screen sizes. Note that, as the wire diameter and mesh size increased, the pressure drop increased in the non-ideal case, since the gas free-flow area decreased.

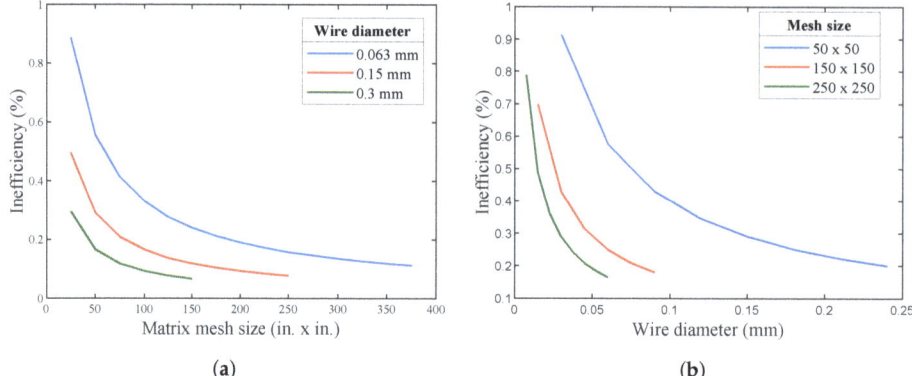

Figure 8. (a) The relation between the matrix mesh size and Ie. (b) The relation between the mesh wire diameter and Ie for several mesh sizes (for a regenerator of 7.5 mm diameter and 30 mm length).

Figure 9a illustrates the relation between the volumetric flow rate on inefficiency for several regenerator diameter to length ratios (D/L). It can be noticed that, for an increased volumetric flow rate, the inefficiency increased due to a decrease in the total number of heat transfer units (NTU$_{Total}$), as illustrated in Figure 9b. In order to maintain the inefficiency at the same level for increased flow rates, the matrix total heat transfer area had to be increased, which meant using a longer regenerator tube with a larger diameter. For the previously highlighted case with a volumetric flow rate of 0.001 m^3/s, NTU$_{Total}$ of 63, $N_x = 5000$, and $N_t = 200$, the inefficiency (Ie) was found to be around 0.21% with $39.8 \leq Re \leq 365$, where Re is the Reynolds' number. This inefficiency level would cause a regenerator thermal loss of ~ 0.39 W, which shows that adequate regenerator performance could still be achieved with such small dimensions.

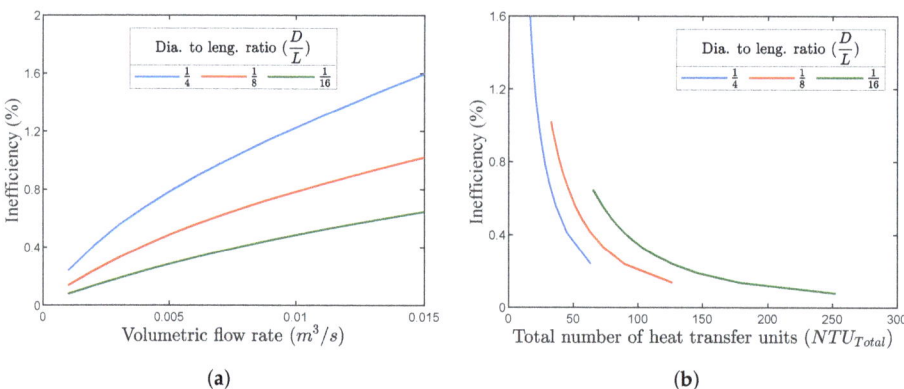

Figure 9. (a) Effect of increasing the volumetric flow rate on the regenerator's inefficiency (Ie) for different regenerator diameter to length ratios. (b) Relation between NTU$_{Total}$ and Ie for the same regenerator diameter to length ratios.

3.2. 2D Regenerator Model

The model assumed that the regenerator temperature was initially at T_h. Figure 10a shows CFD-predicted temperature distributions after four different flow times in the regenerator section. Figure 10b illustrates the regenerator average temperature profile versus flow time, and it shows the temperature dynamics during the heating and cooling periods as the cold and hot gas are flowing in and out of the regenerator. Moreover, it suggests a steady uniform temperature distribution after approximately 80 s in which an average temperature of 195 K was reached. The contour plot in Figure 10a proves this uniform distribution after a flow time of 80 s.

Figure 10c depicts the temperature profiles across the regenerator's center axis after different flow times. It can be seen that the temperature became linearly distributed across the length of the regenerator. The temperature profile similarity between this and Figure 6b proved that the ideal regenerator model succeeded in describing the thermal interaction inside the regenerator.

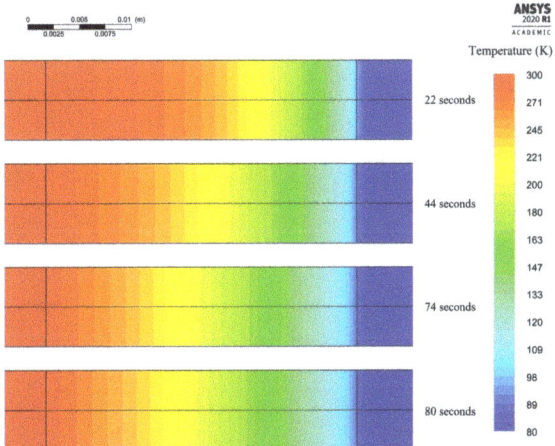

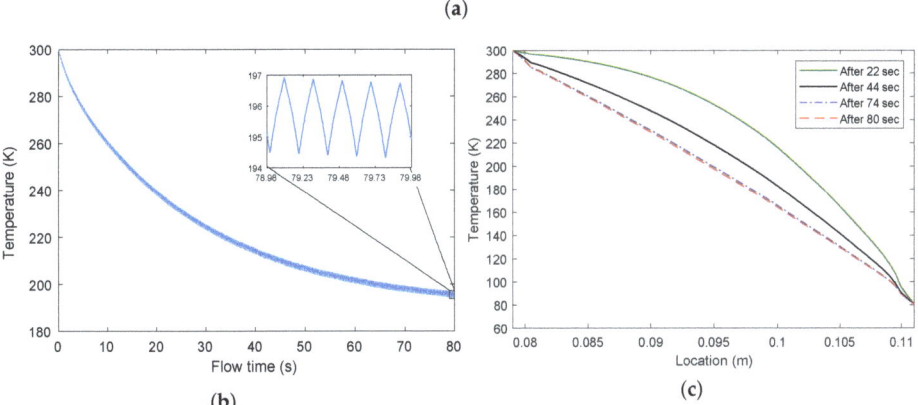

Figure 10. (a) Temperature contours after several flow times. (b) Average temperature versus flow time. (c) Temperature profiles across the regenerator's center axis after different flow times.

Figure 11a,b show the temperature changes, during the heating and cooling periods, at the hot and cold ends, respectively. If a steady flow was assumed, the inefficiency could be calculated using Equations (5) and (6), and these were found to be 2.5% and 2.3% for the cooling and heating periods, respectively. The inefficiency value from the 2D axisymmetric model agreed with the findings in Figure 8b that decreasing the wire diameter would result in decreasing inefficiency. The 2D model proved that the stainless steel #635 mesh size was a good fit for a filler material for a miniature cryogenic regenerator.

Although the ideal regenerator model was a simplified model built on various assumptions and limitations, it not only worked in predicting the thermal behavior inside the regenerator, but also to roughly estimated the effect of the design parameter on the regenerator's performance. However, for more consistent and accurate results, 2D or non-ideal models are required for any design or optimization attempts for the regenerator element. Future work will extend the model to cover the non-ideal case and then compare it to different regenerator CFD models.

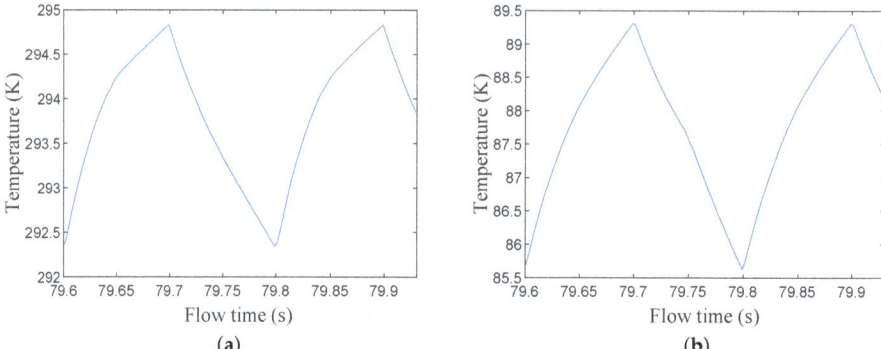

Figure 11. (a) Temperature changes during cooling and heating periods at the regenerator's hot end. (b) Temperature changes during cooling and heating periods at the regenerator's cold end.

4. Conclusions

A MATLAB® code was developed to study the thermal interaction between the working gas and the matrix material in a regenerator element, a component of a closed-cycle regenerative cryocooler. The presented algorithm was a discretization of the ideal regenerator thermal equations. Furthermore, a 2D axisymmetric model was presented to examine the validity of the ideal model and evaluate its findings. This study illustrated the unsteady behavior of the gas and matrix inside the regenerator. It was observed that by increasing the number of spatial and time nodes, the convergence of the algorithm was enhanced. The paper also investigated the effect of regenerator length and diameter on its effectiveness and studied the relation between changing the matrix mesh screen size and its wire diameter on its inefficiency (I_e). It further investigated the implications of changing the volumetric flow rate and the number of heat transfer units on the performance of the regenerator. It was found that these parameters had a significant effect on its inefficiency (I_e) and that an optimum could be sought under given constraints. Finally, the temperature profiles from the 2D model were compared to the ideal model. Future work is to expand the model to consider other substantial losses in the non-ideal case.

Author Contributions: Conceptualization: N.A., J.J.B., and J.G.K.; methodology: N.A.; software: N.A.; validation: N.A.; formal analysis: N.A.; investigation: N.A.; data curation: N.A.; writing, original draft preparation: N.A.; writing, review and editing: J.J.B. and J.G.K.; visualization: N.A.; supervision: J.J.B. and J.G.K.; project administration: J.J.B. and J.G.K. All authors have read and agreed to the published version of the manuscript.

Funding: This research received no external funding.

Acknowledgments: The authors gratefully acknowledge the German Jordanian University (GJU) for partial financial support through a scholarship, as well as the Karlsruhe Institute of Technology (KIT/IMT) for access to laboratories and for an equipment grant. The authors would also like to thank the Editors and the reviewers for their support. We acknowledge support by the KIT Publication Fund of the Karlsruhe Institute of Technology.

Conflicts of Interest: The authors declare no conflict of interest.

References

1. Lewis, R.; Wang, Y.; Cooper, J.; Lin, M.M.; Bright, V.M.; Lee, Y.C.; Bradley, P.E.; Radebaugh, R.; Huber, M.L. Micro cryogenic coolers for IR imaging. *Infrared Technol. Appl. XXXVII* **2011**, *8012*, 80122H.

2. Taekyung, K.; Sangkwon, J. Stirling-type pulse tube refrigerator with slit-type heat exchangers for HTS superconducting motor. *Cryogenics* **2011**, *51*, 341–346.
3. Radebaugh, R. Cryocoolers: the state of the art and recent developments. *J. Phys. Condens. Matter* **2009**, *21*, 164219. [CrossRef] [PubMed]
4. Ackermann, R.A. *Cryogenic Regenerative Heat Exchangers*; Plenum Press: New York, NY, USA, 1997; pp. 147–155.
5. Iliffe, C.E. Thermal analysis of the contra-flow regenerative heat exchanger. *Proc. Inst. Mech. Eng.* **1948**, *159*, 363–372. [CrossRef]
6. Kuriyama, T.; Hakamada, R.; Nakagome, H.; Tokai, Y.; Sahashi, M.; Li, R.; Yoshida, O.; Matsumoto, K.; Hashimoto, T. High efficient two-stage GM refrigerator with magnetic material in the liquid helium temperature region. In *Advances in Cryogenic Engineering*; Springer: Boston, MA, USA, 1990; pp. 1261–1269.
7. Radebaugh, R. A review of pulse tube refrigeration. *Advances in Cryogenic Engineering*; Springer: Boston, MA, USA, 1990; pp. 1191–1205.
8. Kittel, P.; Kashani, A.; Lee, J.M.; Roach, P.R. General pulse tube theory. *Cryogenics* **1996**, *10*, 1191–1205. [CrossRef]
9. Almtireen, N.; Brandner, J.J.; Korvink, J.G. Pulse Tube Cryocooler: Phasor analysis and one dimensional numerical simulation. *J. Low Temp. Phys.* **2020**, 1–19. [CrossRef]
10. Smith, W.R. One-dimensional models for heat and mass transfer in pulse-tube refrigerators. *Cryogenics* **2001**, *41*, 573–582. [CrossRef]
11. Radebaugh, R.; Louie, B. A simple, first step to the optimization of regenerator geometry. In Proceedings of the 3rd Cryocooler Conference, Boulder, CO, USA, 17–18 September 1984; pp. 177–198.
12. Gary, J.M.; Daney, D.E; Radebaugh, R. A computational model for a regenerator. In Proceedings of the 3rd Cryocooler Conference, Boulder, CO, USA, 17–18 September 1984; pp. 199–211.
13. Gary, J.M.; O'Gallagher, A.; Radebaugh, R.; Marquardt, E. *REGEN 3.3: User Manual*; National Institute of Standards and Technology (NIST): Gaithersburg, MD, USA, 2008.
14. Walker, G. *Cryocoolers: Part 2: Applications*; Plenum: New York, NY, USA, 1983.
15. *ANSYS FLUENT 12.0 User's Guide*; Fluent Inc. 2009. Available online: https://www.afs.enea.it/project/neptunius/docs/fluent/html/ug/main_pre.htm (accessed on 17 March 2020).
16. Landrum, E.C. Conrad, T.J. Ghiaasiaan, S.M.; Kirkconnell, C.S. Hydrodynamic parameters of mesh fillers relevant to miniature regenerative cryocoolers. *Cryogenics* **2010**, *50*, 373–380. [CrossRef]

© 2020 by the authors. Licensee MDPI, Basel, Switzerland. This article is an open access article distributed under the terms and conditions of the Creative Commons Attribution (CC BY) license (http://creativecommons.org/licenses/by/4.0/).

Femtosecond Laser-Micromachining of Glass Micro-Chip for High Order Harmonic Generation in Gases

Anna G. Ciriolo [1], Rebeca Martínez Vázquez [1,*], Alice Roversi [2], Aldo Frezzotti [3], Caterina Vozzi [1], Roberto Osellame [1,2] and Salvatore Stagira [1,2]

[1] Institute for Photonics and Nanotechnologies, National Research Council, 20133 Milan, Italy; annagabriella.ciriolo@polimi.it (A.G.C.); caterina.vozzi@polimi.it (C.V.); roberto.osellame@polimi.it (R.O.); salvatore.stagira@polimi.it (S.S.)
[2] Department of Physics, Politecnico di Milano, 20133 Milan, Italy; alice.rov@libero.it
[3] Department of Aerospace Science and Technology, Politecnico di Milano, 20156 Milan, Italy; aldo.frezzotti@polimi.it
* Correspondence: rebeca.martinez@polimi.it

Received: 8 January 2020; Accepted: 1 February 2020; Published: 4 February 2020

Abstract: We report on the application of femtosecond laser micromachining to the fabrication of complex glass microdevices, for high-order harmonic generation in gas. The three-dimensional capabilities and extreme flexibility of femtosecond laser micromachining allow us to achieve accurate control of gas density inside the micrometer interaction channel. This device gives a considerable increase in harmonics' generation efficiency if compared with traditional harmonic generation in gas jets. We propose different chip geometries that allow the control of the gas density and driving field intensity inside the interaction channel to achieve quasi phase-matching conditions in the harmonic generation process. We believe that these glass micro-devices will pave the way to future downscaling of high-order harmonic generation beamlines.

Keywords: femtosecond laser micromachining; high order harmonic generation; de laval gas micro nozzle; attosecond science

1. Introduction

Since its first observation more than twenty years ago, it has been known that an intense and ultrashort laser pulse, focused in a gaseous medium, drives the emission of a burst of coherent radiation, collinear to the driving beam, with a spectral content ranging from the vacuum ultraviolet to the soft X rays [1]. This process is called high-order harmonic generation (HHG) since the spectrum of the emitted radiation appears as a combination of numerous odd harmonics of the fundamental laser field. In the temporal domain, this emission is structured as a train of attosecond light pulses, and indeed, HHG is routinely exploited in the fields of extreme ultraviolet spectroscopy and Attosecond Science [2].

The manipulation of HHG beams is done in grazing incidence on bulky and expensive optics that require careful alignment and even active stabilization systems, and, as a consequence, these beam lines extend over several meters, occupying entire rooms. Up to now, HHG-based eXtreme Ultra Violet (XUV, from 100 nm down to 10 nm) and soft X (from 10 nm down to 1 nm) coherent light sources are confined within a few advanced laboratories [3] because of their technological complexity. In this framework, a substantial breakthrough in ultrafast technology can be achieved by the miniaturization of HHG beamlines, which could remarkably foster the application of HHG sources in numerous novel fields.

A strategy to obtain such compact sources is to generate high-order harmonics in gas-filled glass capillaries of a few centimeters length, which behave as hollow waveguides [4–6]. Hollow waveguides are already widely exploited in the field of ultrafast laser sources mainly for the compression of intense pulses [6], but also in the field of HHG, they lead to impressive results allowing hundreds times enhancement of the intensity of the harmonics to be reached [4] and spectral components up to the keV photon energy [5] in tabletop systems.

A route for achieving even higher miniaturization is based on femtosecond laser micromachining, which is a powerful fabrication technique that has already demonstrated its high potential in the fabrication of miniaturized lab-on-chip devices [7]. Ultrashort laser pulses are focused inside a transparent sample and, due to nonlinear absorption processes, such as multiphoton and avalanche ionization, they produce a permanent modification of the material only at a small region around the focus where the highest intensity is reached [8,9].

The photogenerated hot-electron plasma during fs-laser irradiation induces high temperatures and pressures that give rise to different phenomena, such as densification, direct photo structural modifications, and color-center formation. Under suitable conditions, the combination of such effects may lead to a local increase in the etching speed of the material over a micrometer-sized volume [10]. By moving the laser focus inside the substrate, one can use the laser beam to define three-dimensional regions of increased etching speed. A powerful method for the fabrication of microfluidic networks, in a 3D geometry, directly buried in the glass substrate is femtosecond laser irradiation of fused silica glasses followed by chemical etching (FLICE). Until now, the devices fabricated by FLICE have been extensively used for the manipulation of liquids, but they are perfectly suitable for the manipulation of gases as well.

In this work, we will present gas-filled microfluidic devices, fabricated through the FLICE technique, demonstrating efficient emission of extreme ultraviolet radiation produced by HHG. We exploit the flexibility, accuracy, and 3D capabilities of the FLICE technique to create glass micro-devices for manipulating gas fluxes and for guiding laser beams through microchannels.

2. Materials and Methods

For the femtosecond laser micromachining of the glass devices, the second harmonic (515 nm) of a femtosecond laser beam (Satsuma, Amplitude Systemes S.A., Pessac, France) was focused inside a fused silica sample using a 63 × (0.65 NA) microscope objective (LD-plan Neofluar, Zeiss, Oberkochen, Germany). As represented in the scheme of Figure 1, the glass sample was mounted onto a high-resolution 3D movement system (Aerotech, ANT, Hampshire, UK) and moved with respect to the laser beam following the desired trajectory. The laser repetition rate was set at 1 MHz, with a pulse duration of 230 fs and pulse energy between 200 nJ and 300 nJ depending on the dimensions of the irradiated geometry. After the laser irradiation, the sample was immersed in an ultrasonic bath of a 20% HF aqueous solution at 35 °C.

HHG experiments were performed under vacuum conditions in a beamline composed of an interaction chamber (working pressure 10^{-5} mbar) and a grazing incidence XUV spectrometer (working pressure 10^{-6}–10^{-7} mbar) (see Figure 2). The beam of a Ti:Sapphire laser (25 fs, 1 kHz, 10 mJ) was coupled into the microchannel, which was located inside the interaction chamber. The driving laser beam was focused at the entrance of the hollow waveguide by a 20-cm focal lens. The spatial coupling of the beam with the microchannel was achieved with the help of a high-precision motorized alignment stage. The gas pressure was accurately controlled by a system composed of a needle valve and a pressure gauge placed in the gas line after the needle valve. The high-order harmonics radiation was acquired by a spectrometer composed of grazing-incidence optics. A first toroidal mirror was used for generating an intermediate focus for spectroscopy purposes. The second toroidal mirror focused the HHG beam on a dispersion grating. The dispersed HHG signal was acquired by means of a vacuum compatible detector (Photek, VID140, East Sussex, UK) incorporating one micro-channel plate (MCP) and a P20 phosphor screen. The image displayed on the phosphor screen was acquired by a CCD camera (Apogee Ascent A2150, Andor, Belfast, UK). The same chip was used for HHG for several weeks, with no evident degradation.

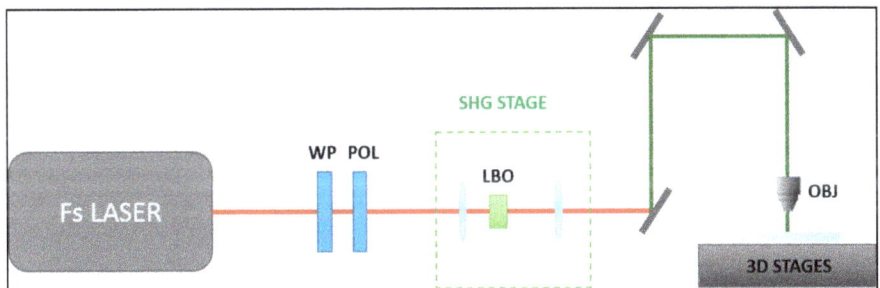

Figure 1. Femtosecond laser micromachining setup. The laser beam passed through an attenuation module, made of a λ/2 waveplate and a polarizer, and then through a telescope and a non-linear crystal (BBO) for second harmonic (515 nm) generation; the second harmonic beam was focused onto the glass sample. The sample was mounted on to a 3-dimension translation stage.

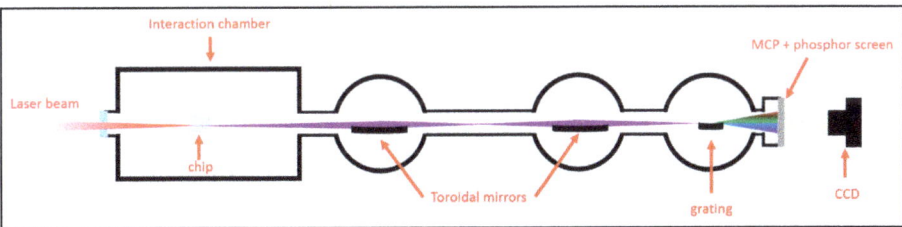

Figure 2. High-order harmonics generation and acquisition setup. The chip was mounted onto a home-built 5-axis translation stage to allow a precise coupling of the driving laser.

We used Comsol Multiphysics software to model the gas speed, temperature, and density (in the steady-state) along the hollow waveguide. We applied the high Mach number flow model coupled with the so-called k-ε turbulence model in order to properly describe the turbulent flow behavior of the gas inside the device. For the discretization of the geometry, we used a triangular mesh, optimized by the software for the physics of the problem.

3. Results and Discussion

The basic design of the glass chip for HHG is shown in Figure 3a; it consisted of a cylindrical top central reservoir (gas pipe) from which several small microchannels (gas distribution channels) departed and reached the cylindrical horizontal main channel (hollow waveguide). The gas was inserted through the top reservoir and, thanks to the homogeneously distributed small microchannels, it uniformly filled the hollow waveguide. In fact, the driving laser that propagated inside the waveguide encountered a uniform gas density.

The final device was made in a fused silica plate (dimensions 6 × 10 × 1 mm^3) and contained two parallel microchannels; the upper one served as an auxiliary channel for beam alignment, whereas the lower one was devoted to HHG. The reservoir for gas input had a radius of 1.6 mm, the small channels for gas distribution had a variable diameter from ~100 μm (at the reservoir base) to ~30 μm at the main channel surface. The main channel, which acted as a hollow waveguide, had a 120 μm diameter. Figure 3b,c show, respectively, the microscope image of the device after laser irradiation and a picture after chemical etching. The in-volume irradiated paths, which look darker in Figure 3b, led to embedded and interconnected empty channels after etching. The irradiated central channel presented a sinusoidal radius to compensate for inhomogeneous exposition to acid during the etching process, and to obtain a central channel with a constant radius after the etching step (see Figure 3c). It is important to point out that the FLICE technique allows combining structures with millimeter size (gas reservoir with 3 mm diameter) and small structures with micrometer size (microchannels) in the same device with just one fabrication process.

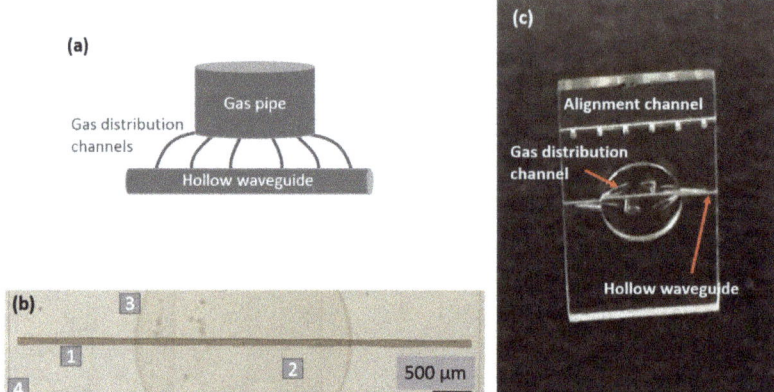

Figure 3. (**a**) Scheme of the high-order harmonic generation (HHG) glass chip, (**b**) microscope image of the irradiated sample, the different components are evidenced: 1-hollow waveguide, 2-gas redistribution channels, 3-gas pipe reservoir, 4-lateral cutting walls, (**c**) picture of the device after the HF etching.

To investigate the gas distribution in the microchip, we modeled the gas flow through the device in the final steady-state using Comsol Multiphysics. The mesh cell dimensions varied between a maximum value of 140 μm and a minimum value of 15 μm. The results, depending on the gas backing pressure, are shown in Figure 4. In particular, the gas temperature, density, and Mach number along the capillary axis are reported. With the exception of small peaks next to the outlets of the distribution channels, the gas flow exhibited the typical features of gas flows within micro-channels with a high length to diameter ratio. The gas flow was nearly isothermal and subsonic in most of the flow region. The density profile smoothly decreased from the center towards the waveguide outlets at the lowest backing pressure value, but it showed stronger spatial modulation at higher pressures. Abrupt changes in the behavior of all flow properties were observed in correspondence to the exits towards the vacuum chamber, where the gas underwent strong expansion.

The gas density inside the channel scaled linearly with the backing pressure. Thus, by properly monitoring the backing pressure, it is possible to achieve accurate control of the density of the generating medium inside the microchannel and, as a consequence, of the generated harmonics intensity.

For the HHG experiments, the gas pipe was connected to the glass chip, which was positioned into the interaction chamber, and a fraction of the laser beam (400 μJ pulse energy) was coupled inside the hollow waveguide. Figure 5 shows a comparison among different harmonic spectra generated inside the microchannel filled with neon gas. A decreasing yield was observed when the gas backing pressure was increased from 50 to 210 mbar. The reduction is more evident for low-order harmonics, as indicated by the reshaping of the harmonic spectrum. This pressure dependence is due to both phase-matching effects and absorption from the gas: as the pressure increases, the XUV radiation is strongly absorbed; moreover, the phase mismatch between the fundamental and the harmonic field worsens and leads to a dramatic reduction in the yield mainly in the low-energy part of the spectrum [11,12].

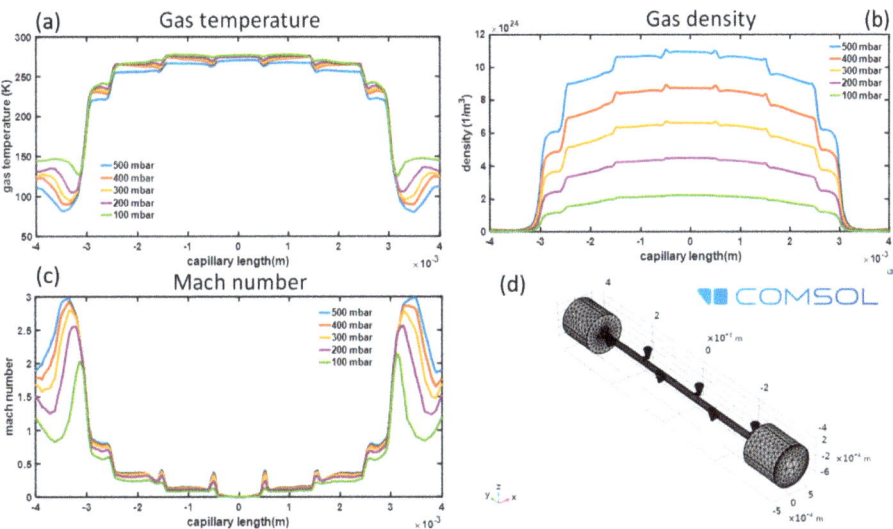

Figure 4. Numerical simulation of the Ne gas evolution inside the micromachined device reported in Figure 2, showing (**a**) gas temperature, (**b**) density, and (**c**) Mach number as a function of the gas backing pressure. (**d**) Gas volume geometry and mesh. The main channel and the small lateral features reproduce the internal chip structure. The two cylindrical structures placed at the extremities were used as output volume for simulating gas expansion in the interaction chamber. The simulations were performed using COMSOL.

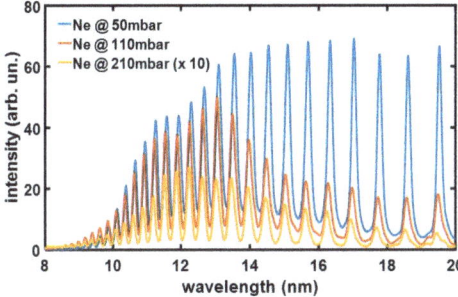

Figure 5. HHG spectra generated inside the chip using Ne gas with a backing pressure of 50 (blue), 110 (red), and 210 mbar (yellow); this latter curve is magnified 10 times.

We compared the optimal HHG yield inside the gas-filled channel with that achieved in the most commonly used interaction geometry based on a gas jet. Figure 6 shows single-shot harmonics spectra generated inside the channel filled with neon at a 50-mbar backing pressure (a) or helium at a 300-mbar backing pressure (b). The results are compared to HHG in a gas jet produced by a 1-mm diameter valve (Parker) that is routinely used for gas-pulse generation in HHG, and that can be operated both in a steady and a pulsed mode. In the steady regime (red spectra), the valve was feed with gas pressures comparable to that used in the channel; in particular, 140 mbar of Neon and 300 mbar of Helium were used. In the pulsed regime (yellow spectra), higher backing pressures, on the multi-bar scale, were used for reproducing the standard working conditions of HHG experiments. Both with Helium and Neon, an extended cutoff, up to 7 nm (160 eV) was obtained inside the microchannel, whereas a cutoff of about 11 nm (110 eV) was observed in the jets. Moreover, an estimation of the harmonic generation yield in the spectral region detected by the spectrometer can be performed by the integration of the HHG spectra. As a result, a higher generation yield was achieved in the microchannel for both gases, which is up to 20 times that achieved with the pulsed jet and about 10^4 times that obtained with the steady jet. The improved performances achieved in the microchannel are related both to the extension of the interaction region and the different interaction geometry with respect to the gas jets.

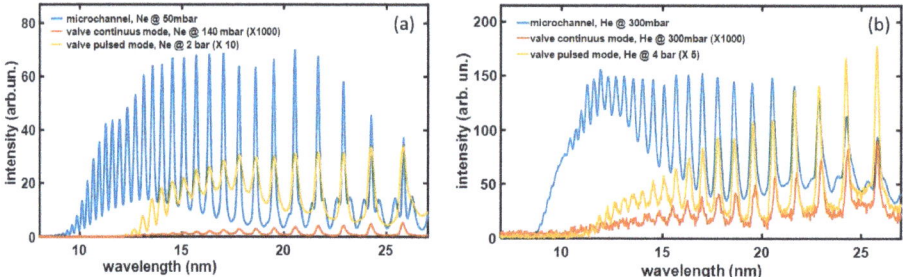

Figure 6. Comparison between HHG spectra produced inside the microchannel (blue), in a continuous gas jet (red), and in a pulsed gas jet (yellow) for Ne (**a**) and He (**b**) gases.

Although there is an overall improvement in the generation yield with respect to traditional HHG in gas, we are confident that a further improvement in the harmonic generation can be obtained by overcoming the phase-matching limitations. In fact, as in many nonlinear processes, HHG efficiency depends both on the single-atom response [13] and on the macroscopic response during propagation in the medium [14], so for generating a bright output beam the emission of a large number of atoms over an extended region must add in phase. The phase relationship between the fundamental and the harmonics field depends on several factors that are mainly related to the generating medium and generation geometry. In particular, the gas neutral and free-electron plasma optical dispersion is wavelength-dependent, so that the fundamental field and the newly generated harmonics can accumulate a phase-mismatch during the propagation. The geometrical phase effects depend instead on the space-structure of the fundamental beam, either tight-focusing or waveguide-assisted confinement.

Practical approaches exploiting quasi phase-matching (QPM) are challenging [5]. In this sense, we propose prototyping different microchips for HHG endowed with multiple gas micro-jets, with the attempt of exploiting QPM. We follow two main strategies which are known to be effective in the improvement in the HHG yield, i.e., the modulation of the driving field intensity or the modulation of the gas density [15,16].

Modulation of the driving field intensity is achievable through modulation of the hollow waveguide diameter. In fact, the diameter of the propagating mode will depend on the waveguide diameter and as a consequence, also the peak intensity [17]. In Figure 7, panels (a) and (b) show the scheme and microscope image of the microchip with the modulated radius. The gas delivery inside the channel is similar to the one in the previous device; it is inserted through the top reservoir, flows through a network of microchannels, and reaches the hollow waveguides at some specific positions. In addition, in this device, each delivering microchannel faces a small exit channel in order to create a modulation of gas density. The strong capabilities of the FLICE technique are evidenced in this glass, which has a 20% radius modulation (110–135 µm) and 30 small gas inlets and outlets with micrometer size.

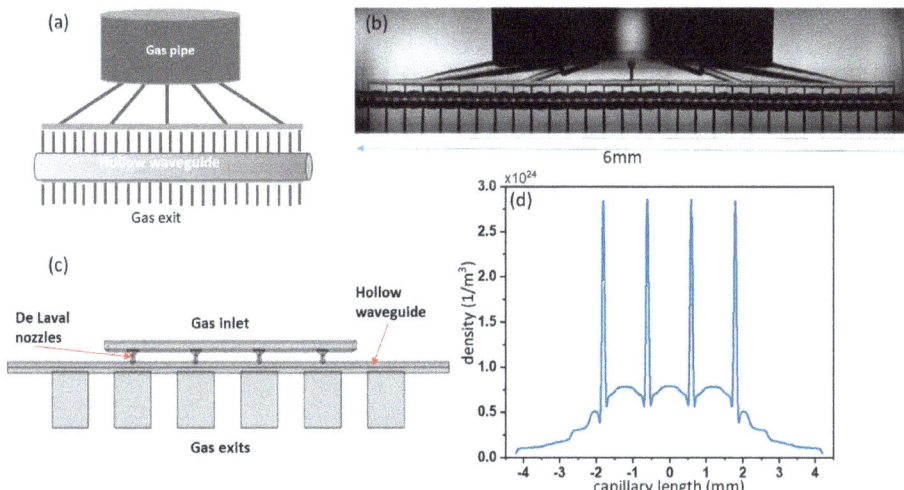

Figure 7. (a) Cartoon of the quasi phase-matching device with 32 gas inlets and modulated diameter. (b) Microscope image of the device in (a). (c) Cartoon of the device with four de Laval micro-nozzles for gas density periodic modulation. (d) Numerical simulation (Comsol) of gas density inside the chip shown in (c).

It is possible to achieve an even higher modulation of the gas density inside the hollow waveguide by injecting the gas through a series of de Laval micro-nozzles whose supersonic flow condition, at the end of the divergent section, helps produce a more focused gas jet. The scheme of the chip is shown in Figure 7c. In this device, there is no gas reservoir, and the convergent section of each nozzle is almost directly facing the gas inlet. Different nozzle geometries were numerically investigated, using Comsol software, in order to enhance/control the density modulation along the axis of the hollow waveguide.

It should be noticed that the nozzle flow behavior in the considered conditions deviates from the classical one, as described by the inviscid theory. Although the inlet pressure is high enough to avoid rarefaction effects, viscous forces are very pronounced (the nozzle throat Reynold's number is about 300) and delay the supersonic transition within the divergent section whose effective area is strongly reduced by a thick boundary layer [18]. So far, the best nozzle configuration consists of a simple double cone geometry with an inlet diameter of 220 µm, a 60 µm throat diameter, and a 90 µm exit diameter, whose size is the largest as possible which preserves the guiding characteristics of the hollow waveguide. The overall nozzle length is 131 µm. Figure 7d shows the simulated gas density profile along the waveguide axis. In order to perform an accurate discretization of the problem, we used different meshes for the nozzle region, central waveguide, and gas exits. In particular, for the micro-nozzles, we used mesh cell dimensions that varied between a maximum element dimension of 5 µm and a minimum element dimension of 0.08 µm (near the walls). A strong density increase in correspondence with the nozzle position is clearly visible (~2 × 10^{24} molecules/m^3). This density modulation is expected to be high enough to obtain QPM, given that the nozzle periodicity is the correct one.

4. Conclusions

In the present work, we demonstrated the application of femtosecond laser micromachined glass microchips to the generation of high-order harmonics in gas. The working principle relays in a main cylindrical microchannel that acts as a hollow waveguide where the driving laser propagates. This microchannel is filled with gas through a three-dimensional network of gas distribution channels all embedded in the glass device. The gas flow was numerically studied and characterized. As a result of the laser-gas interaction, an increase in the HHG signal was observed, which surpasses the performances of the standard generation configuration based on pulsed gas-jets both in terms of efficiency and spectral extension.

By exploiting the extreme flexibility of the FLICE technique, different generation prototypes, including gas-micro-jets and periodically structured capillaries, were realized, aiming at further improving the harmonics generation yield in a quasi phase-matching regime.

We foresee that the FLICE technique could be exploited to downscale an entire HHG-based beamline to a glass chip, integrating several additional functionalities, such as the separation of the IR driving beam from the HHG radiation by waveguide effects, and the inclusion of an interaction module for HHG spectroscopy in liquid and gaseous samples.

Author Contributions: Conceptualization, C.V., R.O., and S.S.; Formal analysis, A.G.C., R.M.V., and A.F.; Funding acquisition, C.V. and S.S.; Investigation, A.G.C., R.M.V., and A.R.; Methodology, A.G.C., R.M.V., and A.F.; Validation, A.G.C., R.M.V., and S.S.; Writing—original draft, A.G.C. and R.M.V; Writing—review and editing, A.G.C., R.M.V., A.F., C.V., R.O., and S.S. All authors have read and agreed to the published version of the manuscript.

Funding: This research was funded by the European Union's Horizon 2020 research and innovation program under the Marie Skłodowska-Curie projects ASPIRE (Grant No. 674960), by the European Research Council Starting Research Grant UDYNI (Grant No. 307964), by the Italian Ministry of Research and Education (ELI project—ESFRI Roadmap).

Conflicts of Interest: The authors declare no conflict of interest.

References

1. Krausz, F.; Ivanov, M. Attosecond physics. *Rev. Mod. Phys.* **2009**, *81*, 163–234. [CrossRef]
2. Calegari, F.; Sansone, G.; Stagira, S.; Vozzi, C.; Nisoli, M. Advances in attosecond science. *J. Phys. B* **2016**, *49*, 62001. [CrossRef]
3. Kühn, S.; Dumergue, M.; Kahaly, S.; Mondal, S.; Füle, M.; Csizmadia, T.; Farkas, B.; Major, B.; Várallyay, Z.; Cormier, E.; et al. The ELI-ALPS facility: The next generation of attosecond sources. *J. Phys. B* **2017**, *50*, 132002. [CrossRef]
4. Tamaki, Y.; Nagata, Y.; Obara, M.; Midorikawa, K. Phase-matched high-order-harmonic generation in a gas-filled hollow fiber. *Phys. Rev. A* **1999**, *59*, 4041–4044. [CrossRef]
5. Popmintchev, T.; Chen, M.-C.; Popmintchev, D.; Arpin, P.; Brown, S.; Alisauskas, S.; Andriukaitis, G.; Balčiūnas, T.; Mucke, O.D.; Pugzlys, A.; et al. Bright Coherent Ultrahigh Harmonics in the keV X-ray Regime from Mid-Infrared Femtosecond Lasers. *Science* **2012**, *336*, 1287–1291. [CrossRef] [PubMed]
6. Nisoli, M.; Stagira, S.; De Silvestri, S.; Svelto, O.; Sartania, S.; Cheng, Z.; Tempea, G.; Spielmann, C.; Krausz, F. Toward a terawatt-scale sub-10-fs laser technology. *IEEE J. Sel. Top. Quantum Electron.* **1998**, *4*, 414–420. [CrossRef]
7. Sima, F.; Sugioka, K.; Vazquez, R.M.; Osellame, R.; Kelemen, L.; Ormos, P. Three-dimensional femtosecond laser processing for lab-on-a-chip applications. *Nanophotonics* **2018**, *7*, 613–634. [CrossRef]
8. Gattass, R.R.; Mazur, E. Femtosecond laser micromachining in transparent materials. *Nat. Photon* **2008**, *2*, 219–225. [CrossRef]
9. Davis, K.M.; Miura, K.; Sugimoto, N.; Hirao, K. Writing waveguides in glass with a femtosecond laser. *Opt. Lett.* **1996**, *21*, 1729–1731. [CrossRef] [PubMed]
10. Marcinkevičius, A.; Juodkazis, S.; Watanabe, M.; Miwa, M.; Matsuo, S.; Misawa, H.; Nishii, J. Femtosecond laser-assisted three-dimensional microfabrication in silica. *Opt. Lett.* **2001**, *26*, 277–279. [CrossRef] [PubMed]
11. Ciriolo, A.G.; Vazquez, R.M.; Crippa, G.; Facciala, D.; Negro, M.; Devetta, M.; Lopes, D.P.; Pusala, A.; Vozzi, C.; Osellame, R.; et al. High-order Harmonic Generation in Femtosecond laser-Micromachined Devices. In Proceedings of the Conference on Lasers and Electro-Optics, San Jose, CA, USA, 13–18 May 2018; p. JTu2A.156.
12. Vázquez, R.M.; Ciriolo, A.G.; Crippa, G.; Vozzi, C.; Stagira, S. Femtosecond laser micromachining of glass chips for high-order harmonic generation (Conference Presentation). *Front. Ultrafast Opt.* **2019**, *10908*, 1090812.
13. Corkum, P.B. Plasma perspective on strong field multiphoton ionization. *Phys. Rev. Lett.* **1993**, *71*, 1994–1997. [CrossRef] [PubMed]
14. Reintjes, J. *Nonlinear Optical Parametric Processes in Liquids and Gases*; Elsevier: Amsterdam, The Netherlands, 2012.

15. Gibson, E.A.; Paul, A.; Wagner, N.; Tobey, R.; Gaudiosi, D.; Backus, S.; Christov, I.P.; Aquila, A.; Gullikson, E.M.; Attwood, D.T.; et al. Coherent Soft X-ray Generation in the Water Window with Quasi-Phase Matching. *Science* **2003**, *302*, 95–98. [CrossRef] [PubMed]
16. Pirri, A.; Corsi, C.; Bellini, M. Enhancing the yield of high-order harmonics with an array of gas jets. *Phys. Rev. A* **2008**, *78*, 011801. [CrossRef]
17. Marcatili, E.A.J.; Schmeltzer, R.A. Hollow Metallic and Dielectric Waveguides for Long Distance Optical Transmission and Lasers. *Bell Syst. Tech. J.* **1964**, *43*, 1783–1809. [CrossRef]
18. Sabouri, M.; Darbandi, M. Numerical study of species separation in rarefied gas mixture flow through micronozzles using DSMC. *Phys. Fluids* **2019**, *31*, 042004. [CrossRef]

© 2020 by the authors. Licensee MDPI, Basel, Switzerland. This article is an open access article distributed under the terms and conditions of the Creative Commons Attribution (CC BY) license (http://creativecommons.org/licenses/by/4.0/).

Article

Optofluidic Formaldehyde Sensing: Towards On-Chip Integration

Daniel Mariuta [1,2], Arumugam Govindaraji [1], Stéphane Colin [2], Christine Barrot [2], Stéphane Le Calvé [3,4], Jan G. Korvink [1], Lucien Baldas [2,*] and Jürgen J. Brandner [1,*]

1. Institute of Microstructure Technology, Campus Nord, Hermann-von-Helmholtz-Platz 1, 76344 Eggenstein-Leopoldshafen, Germany; daniel.mariuta@kit.edu (D.M.); arumugam_g@outlook.com (A.G.); jan.korvink@kit.edu (J.G.K.)
2. Institut Clément Ader (ICA), Université de Toulouse, CNRS, INSA, ISAE-SUPAERO, Mines-Albi, UPS, 31400 Toulouse, France; stephane.colin@insa-toulouse.fr (S.C.); christine.barrot@insa-toulouse.fr (C.B.)
3. Université de Strasbourg, CNRS, ICPEES UMR 7515, F-67000 Strasbourg, France; slecalve@unistra.fr
4. In'Air Solutions, 25 rue Becquerel, 67000 Strasbourg, France
* Correspondence: lucien.baldas@insa-toulouse.fr (L.B.); juergen.brandner@kit.edu (J.J.B.)

Received: 22 May 2020; Accepted: 9 July 2020; Published: 10 July 2020

Abstract: Formaldehyde (HCHO), a chemical compound used in the fabrication process of a broad range of household products, is present indoors as an airborne pollutant due to its high volatility caused by its low boiling point ($T = -19\ °C$). Miniaturization of analytical systems towards palm-held devices has the potential to provide more efficient and more sensitive tools for real-time monitoring of this hazardous air pollutant. This work presents the initial steps and results of the prototyping process towards on-chip integration of HCHO sensing, based on the Hantzsch reaction coupled to the fluorescence optical sensing methodology. This challenge was divided into two individually addressed problems: (1) efficient airborne HCHO trapping into a microfluidic context and (2) 3,5–diacetyl-1,4-dihydrolutidine (DDL) molecular sensing in low interrogation volumes. Part (2) was addressed in this paper by proposing, fabricating, and testing a fluorescence detection system based on an ultra-low light Complementary metal-oxide-semiconductor (CMOS) image sensor. Two three-layer fluidic cell configurations (*quartz*–SU-8–*quartz* and *silicon*–SU-8–*quartz*) were tested, with both possessing a 3.5 µL interrogation volume. Finally, the CMOS-based fluorescence system proved the capability to detect an initial 10 µg/L formaldehyde concentration fully derivatized into DDL for both the quartz and silicon fluidic cells, but with a higher signal-to-noise ratio (*SNR*) for the silicon fluidic cell ($SNR_{silicon} = 6.1$) when compared to the quartz fluidic cell ($SNR_{quartz} = 4.9$). The signal intensity enhancement in the silicon fluidic cell was mainly due to the silicon absorption coefficient at the excitation wavelength, $a(\lambda_{abs} = 420\ nm) = 5 \times 10^4\ cm^{-1}$, which is approximately five times higher than the absorption coefficient at the fluorescence emission wavelength, $a(\lambda_{em} = 515\ nm) = 9.25 \times 10^3\ cm^{-1}$.

Keywords: metal-oxide-semiconductor (CMOS)-based fluorescence sensing; light emitting diode (LED)-induced fluorescence; SU-8 2015 waveguide; silicon fluidic cell; 3,5–diacetyl-1,4-dihydrolutidine (DDL)

---

1. Introduction

Indoor air pollutant concentrations are known to be up to five times higher than outdoors. People generally spend 90% of their time indoors, and approximately 3.8 million people die yearly from diseases related to the exposure to indoor pollution [1]. Among all the indoor pollutants, volatile organic compounds (VOCs) are of particular interest due to their high levels of toxicity [2]. One of the VOCs that raises increased concern is formaldehyde (HCHO), an allergenic, mutagenic, and carcinogenic compound [3,4]. HCHO is largely used in the fabrication process of building materials, household products, and resins for wood products. There are ongoing research projects looking for a non-harmful HCHO alternative in the form of bio-based platform chemical 5-HMF (5-Hydroxymethylfurfural) [5], but meanwhile industrial consumption of HCHO is constantly increasing.

Indoor HCHO concentrations are up to fifteen times higher than those measured outdoors [6] and may have values ranging from 10 to 100 µg/m^3 (1 to 82 parts-per-billion (ppb) at 25 °C and atmospheric pressure) [7]. The World Health Organization (WHO) standard for a safe daily exposure is set to 100 µg/m^3 (82 ppb) concentration for a maximum period of 30 min. In France, the recommendations are to limit the exposure to 30 µg/m^3 (25 ppb), which will be reduced from 2023 onwards to 10 µg/m^3 (8 ppb) [8,9]. The United States Environmental Protect Agency (US EPA) relies on a 2,4-dinitrophenylhydrazine (DNPH) tube sample followed by offline laboratory high-performance liquid chromatography (HPLC) analysis as the standard HCHO detection methodology [10]. This method exhibits the lowest detection limit of 0.048 µg/m^3 (0.04 ppb), but it provides average concentrations over time intervals ranging from one hour to a week for both active and passive samplings, respectively [9,10].

Alternative detection methods have been developed in order to enable real-time and in-field HCHO detection, such as proton transfer reaction mass spectrometry (PTR-MS) [11], gas chromatography mass spectrometry (GC-MS) [12], infrared diode laser spectroscopy [13,14], and Hantzsch reaction monitoring [15]. These methods rely on bulky instruments that are not or barely portable, and are, therefore, not compatible with indoor air monitoring. A device adapted for this need should be real-time, selective, palm-held, low-noise, battery-operated, and cost-effective [7,15].

Among the methods used for online HCHO sensing, the method based on the Hantzsch reaction coupled to fluorescence optical detection (see Figure 1a) is known as the most sensitive and selective [10,15]. The continuous detection methodology can be briefly described as a sequence of four steps: (1) air and reagent streaming; (2) HCHO trapping; (3) derivatization into a fluorescent compound; and (4) optofluidic fluorescence detection [9]. During step (1), the air and reagent phases are continuously put in contact, with the fluids being pressure-driven into the system using micropumps. During step (2), the HCHO molecules from the air are trapped in a 4-amino-3-penten-2-one (Fluoral-P) or acetylacetone solution. The molecular transfer from the gaseous phase to the liquid phase is characterized by a convection–diffusion mechanism [16]. The efficiency of the diffusion process is strictly related to a couple of parameters, such as the contacting area between the two phases, the contacting time, and the molecular driving forces across the gas–liquid interface (e.g., partial pressure difference, chemical potential, concentration difference) [17]. During step (3), a full derivatization process of the HCHO molecule into the 3,5–diacetyl-1,4-dihydrolutidine (DDL) fluorescent molecule takes place for a residence time of $t = 3$ min at $T = 65$ °C (see Figure 1b).

Some works related to on-chip optofluidic integration of HCHO detection in either food [18–20], Chinese herbs [21], or indoor air [7,22] have recently been reported. A microfluidic paper-based analytical device (µPAD) was proposed by Guzman et al. [18] for detection of low HCHO concentrations in food. The detection was made using a complementary metal-oxide-semiconductor (CMOS) camera and the concentration was found by interpreting the images based on RGB color analysis software on a smartphone. The detection mechanism worked offline and the formaldehyde detection range of the device was 0.2–2.5 mg/L. Improvements in terms of portability, response time, and detection range of 0–0.8 mg/L for this system was presented by Liu et al. [19]. Weng et al. [20] integrated four reaction reservoirs and one substrate reservoir on a polydimethylsiloxane (PDMS) microfluidic chip, enabling HCHO detection based on the absorption spectroscopy in 2 µL food samples. A three-layer poly (methyl methacrylate) (PMMA) device with a detection range of 1–50 mg/L was presented by Liu et al. [21] for the detection of HCHO concentrations in Chinese herbs; it was based on laser-induced fluorescence.

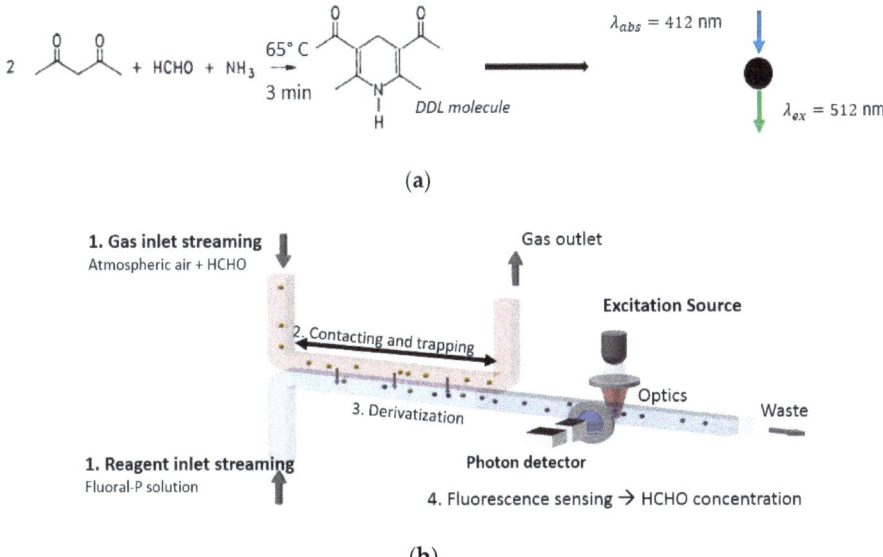

Figure 1. Formaldehyde (HCHO) detection methodology. (**a**) Chemical derivatization reaction of formaldehyde into 3,5–diacetyl-1,4-dihydrolutidine (DDL). (**b**) Gas–liquid contacting, trapping, and detection scheme.

A microfluidic lab-on-a-chip derivatization technique based on the GC-MS measurement technique for HCHO indoor air sensing was described by Pang et al. [22]. The method relied on a glass Pyrex microreactor with a 2.0 m long and 620 µm internal diameter round microchannel as a reactor. The reaction of HCHO with two reagents (pentafluorophenyl hydrazine and O-(2,3,4,5,6-pentafluorobenzyl) hydroxylamine) was studied in a microfluidic context, with the system achieving continuous sampling and analysis with a time resolution of 30 min and a limit of detection (LOD) down to 1.2 µg/m^3 (1 ppb). The reagent flow rates were varied between 20 and 120 µL/min, while the maximum operating temperature and pressure were $T = 300\,°C$ and $p = 3\,MPa$, respectively. Despite the evident advantages of low reactant flow rates, the operating parameters and the time resolution were too high to be compatible with the previously mentioned performances of a HCHO palm-held detector.

Becker et al. [7] developed a microfluidic analytical device (20 cm × 25 cm × 15 cm) dedicated to air analysis based on the Hantzsch reaction, which reached a LOD of 0.13 µg/m^3 (0.1 ppb) for a 17 µL/min reagent flow rate in a microporous tube (10.0 cm length, 0.9 mm internal diameter). The porous tube was placed in a gas microchamber fueled with gaseous HCHO at a flow rate of 250 NmL/min. The temporal resolution was 2 s and the response time was 15 min, enabling near-real-time detection on a portable device. The system performance might be enhanced by simplifying the fluorescence detection system based on a photomultiplier tube (PMT) detector and by reducing the interrogation volume.

Even if the Hantzsch-reaction-based method relies on "wet chemistry", involving a continuous reagent consumption, this method is currently one of the most reliable candidates for developing a sensitive palm-held continuous detector. The alternatives to the wet-chemistry-based methods are electrochemical detection methods. The recently developed electrochemical sensors, despite their rapid response, operate at very high temperatures, are not energy efficient, and cannot be deployed in atmospheres with flammable potential [23]. Their main disadvantages are their low sensitivity and the absence of selectivity; other aldehydes and gas pollutants can interfere with the formaldehyde measurements, as observed by Baldelli et al. [24].

Thus, the aim of developing a palm-held sensitive continuous detector could be achieved by exploiting the recent advancements in the fields of microfluidics, optofluidics (combining microfluidics and optics), microfabrication, and integrated system technology. On-chip integration of microfluidic analytical applications poses multidisciplinary challenges, generally related to: (1) fluid–fluid interactions; (2) fluid–structure interactions; (3) microstructure fabrication; and (4) system integration.

This work presents the preliminary steps and results of the prototyping process towards on-chip integration of an HCHO sensing system. This challenge was split into two individually addressed problems: (1) efficient airborne HCHO trapping in a microfluidic context and (2) low-concentration DDL molecular sensing in reduced interrogation volumes. Problem (1) was partially addressed and a gas–liquid microreactor concept was built around a disposable membrane-based PMMA flat contactor [16]. Problem (2) was addressed by proposing, fabricating, and testing a fluorescence detection system based on an ultrasensitive CMOS image sensor. Two three-layer fluidic cell configurations (*quartz*–SU-8–quartz and *silicon*–SU-8–quartz) were tested, with both possessing a 3.5 µL interrogation volume.

2. Materials and Methods

The fluorescence phenomenon is largely used for molecular optical sensing in microfluidic analytical applications due to its selectivity and capability to reach low detection limits, including single-molecule detection [25,26]. Fluorescence is the property of a molecule allowing it, once excited by light at a specific wavelength λ_{abs}, to emit light after some nanoseconds at a longer wavelength λ_{em}, a behavior known as the Stokes shift. A fluorescent molecule is characterized by the quantum yield ϕ, defined as the ratio of the number of emitted photons to the number of absorbed photons. The number of photons emitted by fluorescence is usually up to three orders of magnitude less than the number of photons being absorbed to excite the molecule [27].

The amount of fluid involved in microfluidic systems is, by definition, reduced. Therefore, the emitted fluorescence signal in optofluidic sensors is very low. Traditional fluorescence detection systems use a complex optical path involving a system of lenses, a high-performance light source, and very sensitive photon detector to capture the low-intensity fluorescence signals [28]. Mainly due to these reasons, the systems are generally expensive, bulky, and barely portable in the best scenarios.

Miniaturization of fluorescence sensing optofluidic devices is a field currently experiencing intense research activity aiming to identify fabrication and design methodologies for an ultraportable, monolithic, and multiplexing sensing architecture. Advances in micro- and nanofabrication technologies have been used to develop new design strategies in order to maximize the detection while shrinking dimensions.

Therefore, when the design of an analytical microsystem based on fluorescence detection is attempted, the Stokes shift and the quantum yield of the molecule are fundamental prerequisites to be considered in order to properly identify the most suitable detection scheme, filtration elements, and light detector. The noise sources in optofluidic fluorescence-based sensors originate from both optical (shot and flicker components) and non-optical elements (dark noise of the detector) [29]. The optical background noise may originate from components with different spectra, e.g., other fluorophores, scattering light, and autofluorescence.

The DDL molecule has two absorption peaks: one in the ultraviolet (UV) range at the 255 nm wavelength and the other in the visible spectrum range at $\lambda_{abs} = 412$ nm, while the fluorescence emission peak occurs at $\lambda_{em} = 515$ nm, with a very low quantum yield $\phi(20\ °C) = 0.005$ [30]. The shorter wavelength possesses a better fluorescence yield but a higher background signal. The absorption peak located at $\lambda_{abs} = 412$ nm was chosen for this prototype, since the UV light sources are more health damaging and less efficient from a cost perspective when compared to the violet light sources.

2.1. Complementary Metal-Oxide-Semiconductor (CMOS)-Based Fluorescence Detector: Sensor Design and Fabrication

The concept of the optical detection system proposed in this work for the detection of the DDL molecules combined a couple of principles usually used in optofluidic sensing microsystems (see Figure 2). The design was based on an orthogonal detection scheme with a CMOS image sensor as the photon detector and high-power light emitting diodes (HP-LEDs) as excitation sources. A square band pass filter was selected and placed in between the fluidic cell and the CMOS image sensor. The components of both optical and fluidic circuits were hosted by two 3D-printed (Ultimaker 3 printer, Utrecht, The Netherlands) holders, named the upper and lower holders.

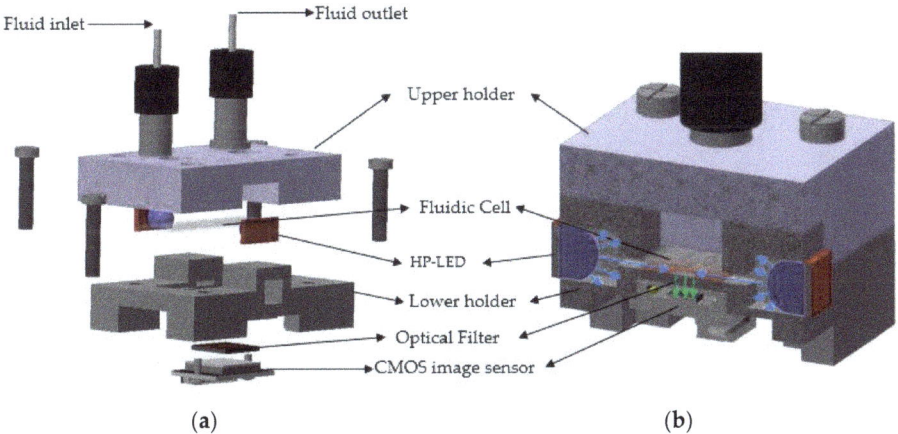

Figure 2. The concept of the complementary metal-oxide-semiconductor (CMOS)-based fluorescence detector (46 mm × 30 mm × 10 mm). (**a**) Exploded view of the detector. (**b**) Transversal cross-section of the assembled sensing device.

The particularity of this design relied on the orthogonal illumination scheme implemented in combination with the contact sensing principle. The contact sensing principle involves the photon detector, the filtration component (if it exists), and the fluidic cell being in direct contact. This is advantageous, since the optical path from the fluorescence source up to the photon detector is diminished, i.e., the optical loss is lower.

2.1.1. Microfluidic Circuit

The fluidic system comprised the fluidic cell, the N-333 type connectors (IDEX Health and Science LLC, Rohnert Park, CA, USA), and the fluidic capillaries. The connectors were integrated into the upper holder. When the upper and lower holders were held together through the four screws, a leakage-free fluidic circuit was obtained, with an important role being played by the rubber sealing of the connectors.

The fluidic cell, i.e., the element of the system hosting the microfluidic flow and the detection chamber from where the fluorescence signal is collected, was designed as a three-layer structure (Figure 3c) and fabricated using well-established photolithography techniques (Figure 3b,d,e).

Two fluidic cell configurations were fabricated, with the difference between the two consisting in the nature of the material of the upper layer. For the first configuration a quartz–SU-8 2015 negative photoresist–quartz structure (quartz fluidic cell) was employed, while for the second configuration a silicon–SU-8 2015 negative photoresist–quartz structure (silicon fluidic cell) was employed.

Quartz was selected due to its low autofluorescence emissions [31], this material being largely used for the fabrication of commercial flow-through fluidic cells. Silicon, a material largely used in the fabrication process of chip technology, possesses interesting properties for the particular aim of this project. More precisely, the absorption coefficient of silicon at the LED emission wavelength, $a(\lambda_{abs} = 420 \text{ nm}) = 5 \times 10^4 \text{ cm}^{-1}$, is approximately five times higher than the absorption coefficient at the emission wavelength, $a(\lambda_{em} = 515 \text{ nm}) = 9.25 \times 10^3 \text{ cm}^{-1}$ [32]. Therefore, the absorption of the excitation light in silicon, which contributes to the optical background noise, is expected to be larger than the absorption of the fluorescence emission, producing an enhancement of the detection limit of the system.

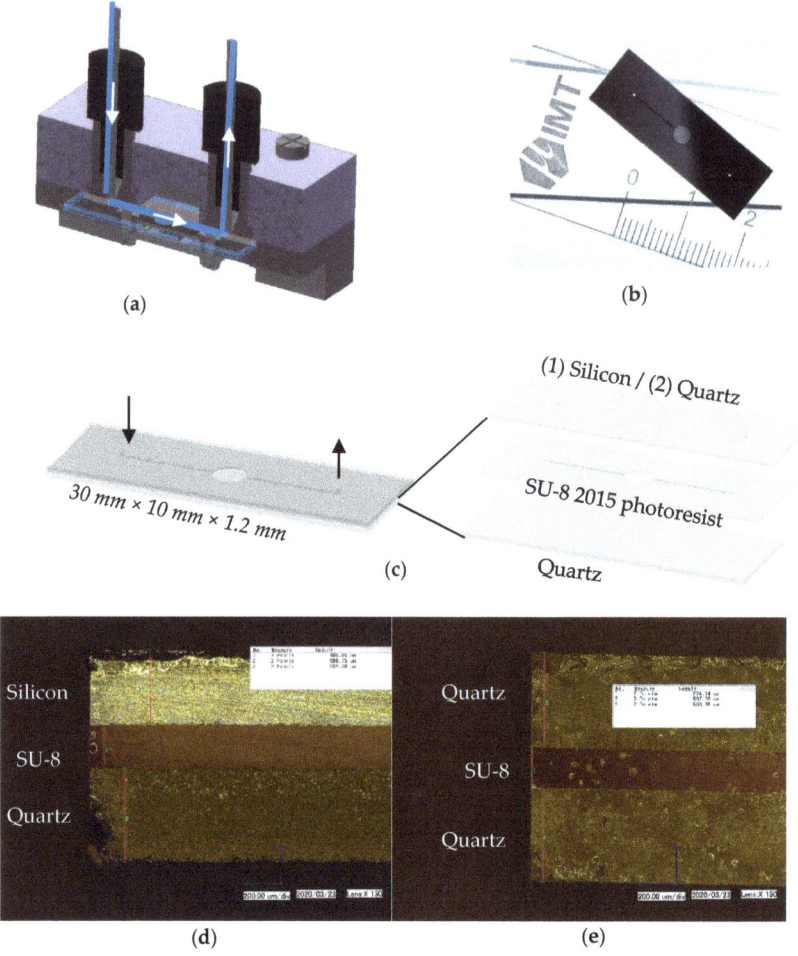

Figure 3. (a) Microfluidic circuit of the detection system. (b) Fabricated silicon fluidic cell. (c) Fluidic cell concept. (d) Microscope view of silicon (405 µm)–SU-8 (232 µm)–quartz (508 µm) fluidic cell cross-section. (e) Microscope view of quartz (507 µm)–SU-8 (224 µm)–quartz (508 µm) fluidic cell cross-section.

The fluidic cell (Figure 3b,d,e) was fabricated at the Laboratoire d'Analyse et d'Architecture des Systèmes (LAAS), Toulouse, France. A 4 inch 500 µm AF32 quartz wafer (Schott AG, Mainz, Germany) was initially cleaned with oxygen plasma at power $P = 800$ W for $t = 5$ min, followed by the deposition of a 200 µm SU-8 2015 negative photoresist layer (Kayaku Advanced Materials, Westborough, MA, USA) and baking using an EVG 120 machine (EVG, St. Florian am Inn, Austria). Afterwards, a mask was placed on top of the SU-8 layer, allowing UV light exposure using a Suss MA6 gen4 instrument (SÜSS MICROTEC SE, Garching, Germany) ($\lambda = 365$ nm, $t = 42$ s, power density $P_d = 14$ mW/cm^2)

in order to create the desired channeling geometry, with a uniform 200 µm depth. Next, the structure was washed using an SU-8 developer to remove the exposed parts, then hard-baked ($T = 125\,°C$, $t = 1$ min). The fluidic inlet and outlet were created through a piercing process on a second 500 µm quartz wafer. Initially, the quartz wafer was coated with Photec 2040 dry film (Hitachi Chemical, Tokyo, Japan) to protect the structure during the piercing process. The piercing was done by sandblasting, followed by rinsing and cleaning with acetone and deionized water. Then, the same oxygen plasma ($P = 800$ W, $t = 5$ min) cleaning procedure used for the first quartz wafer was repeated for the second one. An additional 10 µm SU-8 2015 negative photoresist layer was required on one side of the wafer to facilitate the bonding with the SU-8 2015 layer from the first wafer. Once this layer was coated, the two wafers were bonded using Nanonex NX Nanoimprint equipment (Nanonex, Monmouth Junction, NJ, USA) by applying uniform pressure on both sides. A dicing protection layer made of AZ 4562 photoresist layer (Microchemicals GmbH, Ulm, Germany) was coated and six fluidic cells were finally obtained with different interrogation volumes, as described by Mariuta et al. [33]. The geometry of the detection chamber (Figure 3c) was optimized using a computational fluid dynamics (CFD) model in Ansys Fluent to avoid fluid stagnation regions or dead volumes [33] and to match the sensing area (4.8×4.8 mm^2) of the photon detector.

2.1.2. Optical Circuit

The optical circuit of the sensing system relies on (see Figure 4): (1) the excitation light source; (2) the waveguide (SU-8 2015 negative photoresist layer); (3) the band pass filter; and (4) the CMOS image sensor.

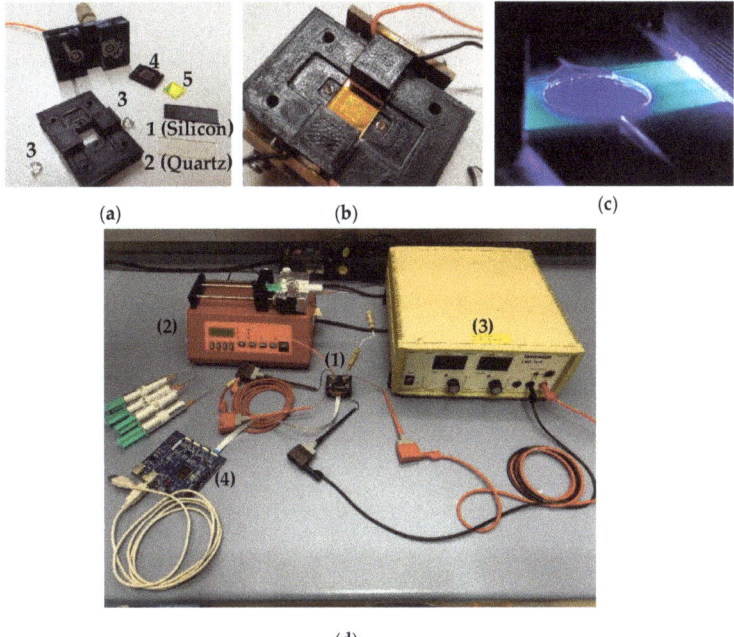

Figure 4. (a) Exploded view of the detection system before assemblage: (1) silicon fluidic cell; (2) quartz fluidic cell; (3) 420 nm light emitting diodes (LEDs); (4) CMOS image sensor; (5) band pass filter. (b) Assembled device;.(c) Light transmission through the fluidic cell towards the detection chamber. (d) (1) Experimental setup: assembled device, (2) syringe pump, (3) 24 V DC power supply, (4) microcontroller board.

LEDs are currently the most widespread excitation choice in microfluidic chemical sensors [34]. The current LEDs available on the market are high-power devices and are relatively cost-efficient. The main disadvantage of the LEDs when they are considered in optical sensing applications is related to the incoherence of the light beam, for which lasers perform better. LEDs are commercially available in the 210–3800 nm spectral range, with their emission spectrum being compatible with the molecules' fluorescence excitation bands [34]. A SMB1N-420H-02 LED (Roithner Lasertechnik GmbH, Vienna, Austria) was selected for this prototype, with a maximal optical radiation power P_{LED} = 420 mW at λ_{LED} = 420 nm emission wavelength, viewing angle of 22°, with the maximum intensity concentrated at the center of its dispersion range. Two LEDs illuminated the detection chamber (see Figure 4a,b), considering its high diameter-over-depth ratio (23.5), to make sure that the excitation photons entirely penetrated the bulk of the interrogation volume.

Photomultiplier tubes (PMT), silicon photodiodes (SPD), CMOS image sensors, and more recently organic photodiodes (OPD) are photon detectors used in the miniaturization of fluorescence-based sensing devices. The recent evolution of the CMOS technology has introduced advantages, such as reduced dimensions, high sensitivity, coupling with filtration algorithms, very low prices per unit, low power consumption, and integrated signal processing, making it compatible for the miniaturization of fluorescence detection. On-chip CMOS integration, coupled with microfluidics for detection at the microscale, has increasingly been investigated in the recent literature [28,35–37], especially for the development of filterless prototypes.

A ULS24 (Anitoa Systems LLC, Menlo Park, CA, USA) CMOS image sensor model was selected as the photon detector for the current prototype, due to its ultra-low light sensitivity (3×10^{-4} lux), low-power supply (3.3 V/1.8 V, 30 mW), maximum sensitivity (60%) at λ_{em} = 515 nm, and minimum sensitivity (40%) at λ_{abs} = 420 nm. The CMOS image sensor was fabricated on a 0.18 µm CIS instrument with a die size of 4.8 × 4.8 mm². The implemented "intelligent dark current management" algorithm made this image sensor capable of matching the sensitivity of the cooled charged-coupled device (CCD) and PMT technologies currently available on the market [38].

The transmission efficiency of the 540 ± 60 nm band pass filter, model BP525-R10, 1 mm thickness (Midwest Optical Systems, Palatine, IL, USA), was tested using a Perkinelmer Lambda 950 UV/VIS spectrometer (PerkinElmer, Waltham, MA, USA). It was found that the transmission efficiency was 0.01% at the excitation wavelength and 91.35% at the emission wavelength (see Figure 5). On top of the band pass filter, was placed a thin metallic layer (200 µm) with a central circular slit of 4.8 mm diameter in order to further minimize the amount of light and then reduce the optical noise induced by the excitation light reaching the CMOS photon detector.

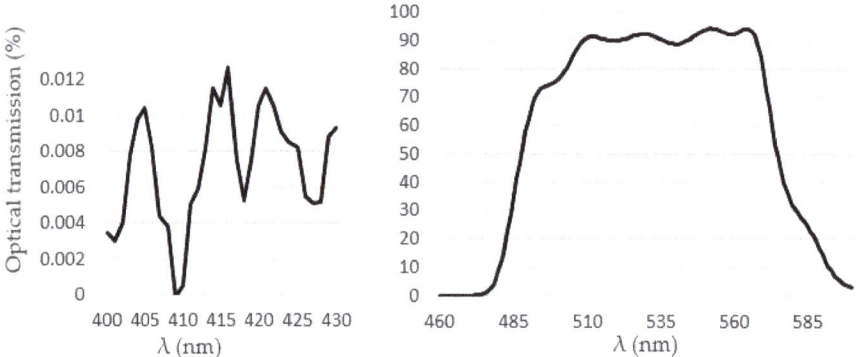

Figure 5. Optical transmittance of the band pass filter.

2.2. CMOS-Based Fluorescence Detector: Experimental Set-Up

2.2.1. Chemicals and Reagents

The Fluoral-P (0.01 M) was prepared by mixing 0.3 mL of acetic acid 100% (Merck, Molsheim, France), 0.2 mL of acetylacetone 99% (Merck), and 15.4 g of ammonium acetate 98% (Sigma-Aldrich, Lyon, France) in 200 mL 18.2 MΩ cm Milli-Q water at 25 °C (Millipore, Molsheim, France). The formaldehyde solutions (c_1 = 0.02 mg/L, c_2 = 0.2 mg/L, c_3 = 2 mg/L, c_4 = 20 mg/L) were prepared at the Institute of Chemistry and Processes for Energy, Environment and Health (ICPEES) Strasbourg by mixing a commercial formaldehyde solution (37% in water, Sigma-Aldrich) with Mili-Q water (18.2 MΩ·cm at 25 °C, Millipore). DDL solutions containing initial formaldehyde concentrations (c_1 = 0.01 mg/L, c_2 = 0.1 mg/L, c_3 = 1.0 mg/L, c_4 = 10.0 mg/L) were prepared just before the measurements by mixing Fluoral-P (0.02 M) and formaldehyde solutions (1:1 v/v), then placing it these an oven at T = 65 °C for t = 3 min to fully derivatize formaldehyde into DDL.

2.2.2. Experimental Procedure

The two LEDs were connected in series, with the power being supplied from a 24 V DC source. The LEDs were turned on for short periods ($t \cong 1$ s), while the data from the CMOS image sensor were captured using a PC interface during this activation time of the excitation source. Heat can cause a shift in the LED emission band, reducing the detection accuracy. Thus, the illumination of the detection chamber was done in short pulses ($t \cong 1$ s) in order to avoid a temperature increase of the LEDs, which were provided with flat copper heat sinks. The interface provided a 12 × 12 matrix, representing the photon counts measured by each pixel, the temperature of the CMOS sensor at the measurement time, and a mean value $\overline{X}$ representing the average of all the X_i 12 × 12 photon counts.

The ULS 24 CMOS image sensor was connected to the microcontroller board (Anitoa Systems LLC, Menlo Park, CA, USA), then to a personal computer (Figure 4d). The CMOS image sensor was settled to work in low-gain mode, 12 × 12-pixel configuration, and t = 1 ms integration time.

The response of the system was iteratively tested for different DDL solutions at various formaldehyde concentrations (c_1 = 0.01 mg/L, c_2 = 0.1 mg/L, c_3 = 1.0 mg/L, c_4 = 10.0 mg/L, and for the blank concentration c_0 = 0.0 mg/L). During the measurement process, three different optical powers of the LED were employed: $P_1 = 0.25\ P_{LED} = 105$ mW, $P_2 = 0.5\ P_{LED} = 210$ mW, $P_3 = 0.75\ P_{LED} = 315$ mW. The DDL solutions were continuously sampled during the measurements at a flow rate of 10 μL/min.

3. Results and Discussion

The experimental response of the system was evaluated by applying a regression analysis technique to the experimental data obtained from two independently prepared DDL samples (N = 2, sample #1 and sample #2) for each of the tested concentrations, $c_{i,\ i\ =\ 0,...,\ 4}$. Each of the samples processed in the system was analyzed by taking seven continuous measurements (n = 7) of the photon intensities, represented by $X_{i,\ i\ =\ 1,...,\ 7}$ in Equation (1). X_i was in fact the mean value of all photon intensities given by the 12 × 12 pixel matrix obtained at the particular measurement time. Here, $\overline{X}_{\#1}$ represented the mean value of the $X_{i,\ i\ =\ 1,...,\ 7}$ measurements taken for the first sample, while $\overline{X}_{\#2}$ represented the mean value for the second sample investigated at a specific formaldehyde concentration of DDL solution (see Tables 1 and 2).

The precision of the evaluation was quantified by calculating the absolute (σ) (Equation (1)) and relative standard deviations (RSD) (Equation (2)), and the performance was estimated using the signal-to-noise ratio (SNR) (Equation (3)). The SNR was used to describe the signal increase relative to the blank sample response (c_0 = 0.0 mg/L), but its relevance was limited due to the low number of independent samples tested. While $\sigma_{\#1}$ and $\sigma_{\#2}$ represented the standard deviations obtained for measurements taken for similar samples ($\overline{X}_{\#1}$ and $\overline{X}_{\#2}$), σ represented the standard deviation of $\overline{X}_{N=2} = (\overline{X}_{\#1} + \overline{X}_{\#2})/2$ values.

$$\sigma = \sqrt{\frac{\sum_{i=1}^{n}(X_i - \overline{X})}{N}}, \text{ with } \overline{X} = \frac{1}{n}\sum_{i=1}^{n} X_i \tag{1}$$

$$RSD = \frac{\sigma}{\overline{X}_{N=2}} \tag{2}$$

$$SNR(c_i) = \frac{\overline{X}_{N=2}(c_i) - \overline{X}(c_0)}{\sigma(c_i)} \tag{3}$$

Memory effects (see Figure 6) were observed during the initial stages of the experimental campaign. Therefore, the fluidic cells were rinsed with acetone and dried in the oven after each sampling.

Table 1. Error distribution as the standard deviation (σ), relative standard deviation (RSD), and the signal-to-noise ratio (SNR) for the quartz fluidic cell at the excitation optical power, $P = 315$ mW, where #1 and #2 represented the two independently prepared samples for each of the tested formaldehyde concentrations c_i for $i = 0, \ldots, 4$.

Formaldehyde Concentration (mg/L)	0.00	0.01	0.10	1.0	10.0
$\overline{X}_{\#1}$ (counts)	1122	1477	1505	1528	1611
$\sigma_{\#1}$	3	2	3	4	5
$RSD_{\#1}$ (%)	0.21	0.16	0.21	0.29	0.29
$\overline{X}_{\#2}$ (counts)	1155	1230	1323	1489	1546
$\sigma_{\#2}$	3	9	2	1	3
$RSD_{\#2}$ (%)	0.24	0.71	0.16	0.06	0.20
$\overline{X}_{N=2}$ (counts)	1189	1354	1414	1509	1578
$\overline{X}_{N=2}$	34	123	91	20	33
RSD (%)	2.82	9.11	6.43	1.29	2.06
SNR	-	4.9	6.7	9.6	11.6

Table 2. Error distribution as the standard deviation (σ), relative standard deviation (RSD), and signal-to-noise ratio (SNR) for the silicon fluidic cell at the excitation optical power $P = 315$ mW, where #1 and #2 represented the two independently prepared samples for each of the tested concentrations c_i for $i = 0, \ldots, 4$.

Formaldehyde Concentration (mg/L)	0.00	0.01	0.1	1.0	10
$\overline{X}_1$ (counts)	771	902	988	966	1,018
$\sigma_{\#1}$	3	1	2	15	2
$RSD_{\#1}$ (%)	0.36	0.15	0.24	1.59	0.22
$\overline{X}_2$ (counts)	822	1005	1066	1053	1184
$\sigma_{\#2}$	4	8	2	2	1
$RSD_{\#2}$ (%)	0.44	0.75	0.21	0.16	0.09
$\overline{X}_{N=2}$ (counts)	797	954	1027	1010	1101
$\overline{X}_{N=2}$	26	51	39	43	83
RSD (%)	3.21	5.39	3.82	4.27	7.54
SNR	-	6.1	9.0	8.3	11.9

The response of the CMOS-based fluorescence detector for a blank sample ($c_0 = 0.00$ mg/L) was plotted in Figure 7b. By increasing the excitation optical power, a perfect linear response ($R2 = 0.999$) was obtained for both quartz ($y = 351.5\,x + 141$) and silicon ($y = 226.5\,x + 120.3$) cells. However, a higher steepness of the linear equation was observed for the quartz fluidic cell compared to the silicon fluidic cell. Moreover, lower values of the photon counts were measured for the silicon cell. The lower intensity and steepness measured for the silicon cell may have been caused by two main factors: (1)

lower excitation light intensity reaching the detection chamber, and consequently the photon detector, due to the 400 μm thickness of the opaque silicon upper layer; (2) the light absorbance of silicon at λ_{abs} = 420 nm is approximately five times higher than the absorbance of silicon at λ_{em} = 515 nm. This latter factor might also explain the differences between the SNRs obtained for the two fluidic cell configurations (see Tables 1 and 2, Figure 7a). Analyzing the results obtained for both fluidic cells (Figure 8b,c), it was observed that the fluorescence intensity measured for the silicon cell was generally lower compared to that of the quartz cell.

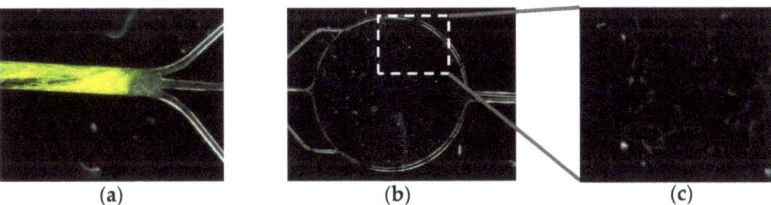

(a) (b) (c)

Figure 6. (a) Dry green DDL crystals cluster after streaming a DDL solution at a formaldehyde concentration of c_5 = 100 mg/L. (b) Detection chamber with (c) green DDL crystals on the inner walls after use of the DDL solution at a formaldehyde concentration of c_4 = 10 mg/L.

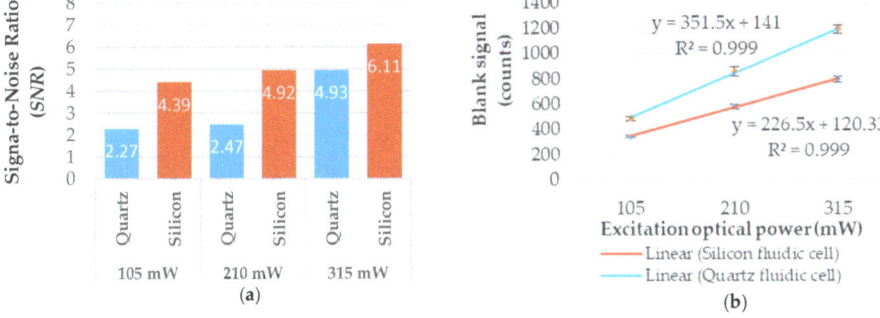

Figure 7. (a) Signal-to-noise ratios (SNRs) obtained for quartz and silicon fluidic cells at a formaldehyde concentration of c_1 = 0.01 mg/L in DDL solution and with different optical powers. (b) The linear response obtained for three excitation optical powers at c_0 = 0.00 mg/L (blank sample). The error bars indicate the standard deviations for two independently prepared samples (N = 2).

From the presented experimental dataset, relatively high values of the photon counts were observed for the blank sample (see Figure 7b). This was mainly due to the fact that the transmission efficiency of the band pass filter at the excitation wavelength was not exactly zero (see Figure 4). This fact caused 0.008% of the excitation optical signal to reach the very sensitive ULS24 CMOS image sensor.

The CMOS image sensor response as a 12 × 12 light intensity matrix representation is illustrated in Figure 8a. The image corresponding to the CMOS dark noise depicted its response in a complete dark environment (no excitation light). The second picture from the left to the right illustrated the CMOS image sensor response for a 0.0 mg/L concentration solution (blank sample), with a value of 1225 counts being registered. As expected, higher values were obtained once the DDL concentration streamed into the system was iteratively increased. Due to the fact that the differences between the pixel outputs were relatively low compared to the dynamic range of the image sensor, the gray color gradient found was also low. However, there was a visible slight enhancement in the gray color intensity from the left to the right coming from the circular detection volume.

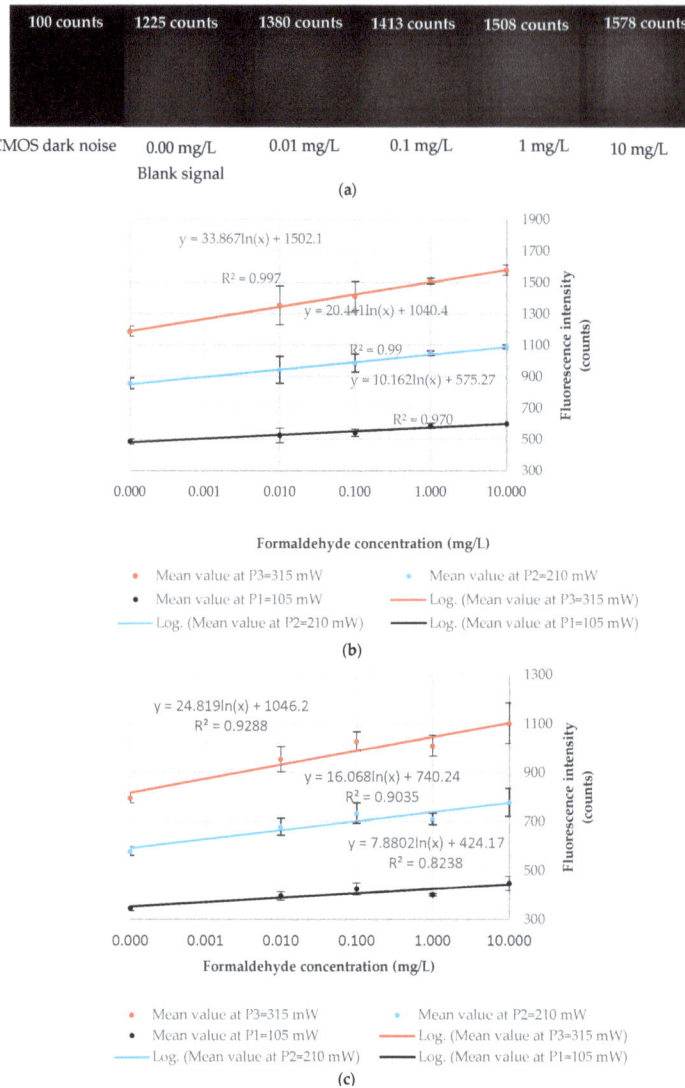

Figure 8. Experimental evaluation of the CMOS-based fluorescence detector realized using the mean values ($\overline{X}$) of seven measurements ($n = 7$) made on two independently prepared samples ($N = 2$) at each of the c_i, $i = 0, \ldots, 4$ concentrations tested. (**a**) ULS 24 CMOS image sensor 12 × 12-pixel output for the dark current, blank signal, and the four formaldehyde concentrations tested for a quartz cell at $P = 315$ mW. (**b**) Logarithmic fitting of the experimental points for a quartz fluidic cell. (**c**) Logarithmic fitting of the experimental points for a silicon fluidic cell. The error bars indicate one standard deviation ($N = 2$).

The experimental response of the CMOS-based fluorescence detector presented in Figure 8b,c shows a logarithmic correlation between the DDL fluorescence intensity and formaldehyde concentrations in DDL solutions. Analyzing the plots, it was observed that the correlation coefficients (R^2) increased proportionally with the excitation optical power for both cells. The logarithmic behavior suggested that the concentration range tested was above the limit of linearity of the system [39]. A logarithmic behavior of the fluorescence signal was observed in the 0.01–10 mg/L formaldehyde concentration range for both cells, indicating that the upper limit of the linear range is most probably below the lowest concentration tested, i.e., 0.01 mg/L. The RSD calculated for measurements performed on a particular sample gave values ranging from 0.06% up to 0.71% for the quartz cell (see Table 1) and values ranging 0.09% to 1.59% for the silicon cell (see Table 2). When RSDs were calculated for the duplicate samples, values ranging from 1.29% to 9.11% for the quartz cell and from 3.82% to 7.54% for the silicon cell were found. The source of errors might be related to the inherent differences that arose during the preparation of the samples, the uncontrolled temperature of the environment, and the low number of replicates. Moreover, bubbles were sometimes observed at the end of the measurement process inside the fluidic cell detection chamber. The observed fluorescence intensity mean values were higher when bubbles were present inside the chamber compared to the situation when bubbles were absent. This is why the measurements made in the presence of air bubbles occupying parts of the detection chamber were not considered when the response of the sensor was determined. The sources of error mentioned above might be the reason why the mean values obtained for the silicon fluidic cell at $c_3 = 1.0$ mg/L were observed as being lower than expected for all the three optical excitation powers tested (see Figure 8c). Analyzing the plot given by the other three concentrations, it could be observed that the values obtained at this concentration were below the trend line; moreover, the values obtained at $c_3 = 1.0$ mg/L were lower than the values obtained at $c_3 = 0.1$ mg/L.

Although the results obtained were approximate, a tendency toward higher SNRs in the case of the silicon fluidic cell could be noted. The SNR obtained for the lower formaldehyde concentration tested, $c_1 = 0.01$ mg/L, is plotted in Figure 7a. At the maximum optical power tested, the SNR value of the quartz cell was found to be $SNR_{quartz} = 4.9$, while $SNR_{silicon} = 6.1$.

4. Conclusions and Perspectives

In conclusion, it should be noted that the main achievements registered in this work were related to the on-chip integration of the detection part of the sensor. A CMOS-based fluorescence detector was developed and proved to be capable of detecting 10 µg/L of formaldehyde derivatized into DDL ($\phi(20\ °C) = 0.005$) in a 3.5 µL interrogation volume with $SNR = 4.9$ and $RSD = 9.1\%$ ($N = 2$) for a quartz fluidic cell; and with $SNR = 6.1$ and $RSD = 5.3\%$ ($N = 2$) for a silicon fluidic cell. Moreover, the enhancement of the signal intensity in the silicon fluidic cell due to its absorption coefficient at the LED emission wavelength, $a(\lambda_{abs} = 420\ nm) = 5 \times 10^4\ cm^{-1}$, which is approximately five times higher than the absorption coefficient at the emission wavelength, $a(\lambda_{em} = 515\ nm) = 9.25 \times 10^3\ cm^{-1}$, was confirmed.

Although the repeatability of the results has to be tested for more independently prepared samples in order to fully validate the system, the potential of the concept has been demonstrated through the obtained results. The logarithmic behavior of the fluorescence signal was observed in the 0.01–10 mg/L formaldehyde concentration range for both cells, indicating that the upper limit of the linear range is most probably below the lowest concentration tested, i.e., 0.01 mg/L. The linear range of the system could be obtained after lowering the optical background noise level, which was observed as being relatively high.

Otherwise, the feasibility of the detection configuration has been proven, with the system showing high miniaturization and detection improvement potential. The 1 mm thickness band pass filter could be replaced by a filtration layer coated directly on top of the CMOS image sensor, as described by Guduru et al. [40] and Sunaga et al. [41]. This would further diminish the optical path length and lead towards a fully-integrated monolithic device. Moreover, a filterless configuration based on time-resolved detection methodology could be employed, as described by Mariuta et al. [16]. This methodology practically avoids the optical background noise, but its implementation is strictly related to the working frequency of the processor involved, since the DDL molecule fluorescence decay time is only limited to 2 ns.

The study showed the functionality of the concept based on an orthogonal detection scheme, employing a LED-induced fluorescence technique coupled to a CMOS image sensor as an ultra-low light detection system. Due to the memory effects that were consistently observed, the fluidic channels were rinsed with acetone after each sampling.

In order to further reduce the fabrication costs, PMMA polymer, providing very good optical properties for this particular application, might be considered as the main material for the fabrication and testing of a new fluidic cell. Its implementation as a waveguide for biomedical applications and its potential for wearable and in vivo applications, which are currently being studied [42], makes it a promising material for such an application. Here, attention has to be paid to the surface roughness that results from the fabrication process, possible swelling from liquids, and its own fluorescence. Another possible direction could be represented by a full silicon structure, considering the CMOS image sensor validation for this application and the SNR enhancement observed for silicon.

In addition, the potential of this device could go beyond this specific application, contributing to the rapidly developing field of on-chip fluorescence sensing for various applications of rapid and low-cost monitoring in chemistry, biology, or environmental fields.

Regarding the challenge related to airborne HCHO trapping in a microfluidic context, a gas–liquid microreactor concept was proposed [16]. A disposable gas–liquid microcontactor, designed as a two-layer PMMA structure with an embedded polymeric membrane, is supposed to be integrated in between two holders equipped with an integrated heating and fluidic streaming system. Each PMMA layer of the contactor hosts a meandering network of rectangular cross-section microchannels, with one layer being assigned to the reagent streaming (liquid carrying layer) and one to the gas streaming (gas carrying layer).

After successful fabrication of the subsystems of the gas–liquid microreactor, further results are expected to experimentally prove the concept. Further analytical studies have to be performed in order to check the viability of the on-chip membrane integration technique and to study the mass transfer efficiency at different flow rates of gas and reagent streams. A mathematical model for the estimation of formaldehyde trapping in the reagent stream through a hydrophobic membrane was also developed and presented by Mariuta et al. [16]. The interest here is to enable enhanced and efficient formaldehyde trapping using cost-efficient on-chip membrane-based polymer chips. Next, a system integration study has to be performed in order to identify solutions for low-noise, cost-effective fluid pumping.

Author Contributions: Conceptualization, D.M., S.L.C., L.B., and J.J.B.; methodology D.M. and A.G.; software, D.M. and A.G.; validation, D.M., A.G., J.J.B., J.G.K., and L.B.; formal analysis, D.M. and J.J.B.; investigation, D.M.; resources, L.B., C.B., S.C., S.L.C., J.J.B., and J.G.K.; data curation, D.M.; writing—original draft preparation, D.M.; writing—review and editing, L.B., C.B., S.C., S.L.C., J.J.B., and J.G.K.; supervision, J.J.B., L.B., J.G.K., and S.L.C.; project administration, J.J.B., L.B., S.C., C.B., and J.G.K.; funding acquisition, J.J.B. and L.B. All authors have read and agreed to the published version of the manuscript.

Funding: The work in this article has received funding from the European Union's Horizon 2020 research and innovation programme under the Marie Sklodowska-Curie grant agreement No 643095.

Acknowledgments: The authors want to acknowledge the contribution of Professor Aldo Frezzotti from Politecnico di Milano, Italy, Christian Karnutsch from Hochschule Karlsruhe, Achim Voigt, and Alexandra Moritz from Institute of Microstructure Technology, Karlsruhe. This work was partly supported by the French Renatech network under the project P-17-02324. We acknowledge support by the KIT-Publication Fund of the Karlsruhe Institute of Technology.

Conflicts of Interest: The authors declare no conflict of interest.

References

1. Household Air Pollution: Pollutants. Available online: https://www.who.int/airpollution/household/pollutants/en/ (accessed on 24 December 2019).
2. Squire, R.A.; Cameron, L.L. An analysis of potential carcinogenic risk from formaldehyde. *Regul. Toxicol. Pharmacol.* **1984**, *4*, 107–129. [CrossRef]

3. Casset, A.; Marchand, C.; Purohit, A.; Le Calve, S.; Uring-Lambert, B.; Donnay, C.; Meyer, P.; de Blay, F. Original article Inhaled formaldehyde exposure: Effect on bronchial response to mite allergen in sensitized asthma patients. *Allergy* **2006**, *61*, 1344–1350. [CrossRef] [PubMed]
4. Nielsen, G.D.; Larsen, S.T.; Wolkoff, P. Re-evaluation of the WHO (2010) formaldehyde indoor air quality guideline for cancer risk assessment. *Arch. Toxicol.* **2017**, *91*, 35–61. [CrossRef]
5. 5-HMF: The Key to Green Chemistry. Available online: https://5-hmf.com/ (accessed on 24 December 2019).
6. Salthammer, T.; Mentese, S.; Marutzky, R. Formaldehyde in the indoor environment. *Chem. Rev.* **2010**, *110*, 2536–2572. [CrossRef]
7. Becker, A.; Andrikopoulou, C.; Bernhardt, P.; Ocampo-Torres, R.; Trocquet, C.; Le Calvé, S. Development and Optimization of an Airborne Formaldehyde Microfluidic Analytical Device Based on Passive Uptake through a Microporous Tube. *Micromachines* **2019**, *10*, 807. [CrossRef]
8. ANSES - ANSES - French Agency for Food, Environmental and Occupational Health & Safety. Indoor Air Quality Guidelines (IAQGs) - Formaldehyde. Available online: https://www.anses.fr/en/content/indoor-air-quality-guidelines-iaqgs (accessed on 9 July 2020).
9. Guglielmino, M.; Allouch, A.; Serra, C.A.; Le Calvé, S. Development of microfluidic analytical method for on-line gaseous Formaldehyde detection. *Sens. Actuators B Chem.* **2017**, *243*, 963–970. [CrossRef]
10. Liu, J.; Li, X.; Yang, Y.; Wang, H.; Kuang, C.; Chen, M.; Hu, J.; Zeng, L.; Zhang, Y. Sensitive Detection of Ambient Formaldehyde by Incoherent Broadband Cavity Enhanced Absorption Spectroscopy Sensitive Detection of Ambient Formaldehyde by Incoherent Broadband Cavity Enhanced Absorption Spectroscopy. *Anal. Chem.* **2020**, *92*, 2697–2705. [CrossRef] [PubMed]
11. Warneke, C.; Veres, P.; Holloway, J.S.; Stutz, J.; Tsai, C.; Alvarez, S.; Rappenglueck, B.; Fehsenfeld, F.C. Airborne formaldehyde measurements using PTR-MS: Calibration, humidity dependence, inter-comparison and initial results. *Atmos. Meas. Tech.* **2011**, *4*, 2345–2358. [CrossRef]
12. Dugheri, S.; Bonari, A.; Pompilio, I.; Colpo, M.; Mucci, N.; Montalti, M.; Arcangeli, G.; Brambilla, L.G.A.; Morgagni, V.G.B. Development of an Innovative Gas Chromatography – Mass Spectrometry Method for Assessment of Formaldehyde in the Workplace Atmosphere. *Acta Chromatogr.* **2017**, *29*, 511–514. [CrossRef]
13. Cihelka, J.; Matulková, I.; Civiš, S. Laser diode photoacoustic and FTIR laser spectroscopy of formaldehyde in the 2.3 µm and 3.5 µm spectral range. *J. Mol. Spectrosc.* **2009**, *256*, 68–74. [CrossRef]
14. Hanoune, B.; LeBris, T.; Allou, L.; Marchand, C.; Le Calvé, S. Formaldehyde measurements in libraries: Comparison between infrared diode laser spectroscopy and a DNPH-derivatization method. *Atmos. Environ.* **2006**, *40*, 5768–5775. [CrossRef]
15. Descamps, M.N.; Bordy, T.; Hue, J.; Mariano, S.; Nonglaton, G.; Schultz, E.; Tran-Thi, T.H.; Vignoud-Despond, S. Real-time detection of formaldehyde by a sensor. *Sens. Actuators B Chem.* **2012**, *170*, 104–108. [CrossRef]
16. Mariuta, D.; Baldas, L.; Colin, S.; Le Calvé, S.; Korvink, J.G.; Brandner, J.J. Prototyping a Microfluidic Sensor for Real - Time Detection of Airborne Formaldehyde. *Int. J. Chem. Eng. Appl.* **2020**, *11*, 23–28.
17. Wibisono, Y.; Cornelissen, E.R.; Kemperman, A.J.B.; van der Meer, W.G.J.; Nijmeijer, K. Two-phase flow in membrane processes: A technology with a future. *J. Memb. Sci.* **2014**, *453*, 566–602. [CrossRef]
18. Guzman, J.M.C.C.; Tayo, L.L.; Liu, C.C.; Wang, Y.N.; Fu, L.M. Rapid microfluidic paper-based platform for low concentration formaldehyde detection. *Sens. Actuators B Chem.* **2018**, *255*, 3623–3629. [CrossRef]
19. Liu, C.C.; Wang, Y.N.; Fu, L.M.; Huang, Y.H. Microfluidic paper-based chip platform for formaldehyde concentration detection. *Chem. Eng. J.* **2018**, *332*, 695–701. [CrossRef]
20. Weng, X.; Chon, C.H.; Jiang, H.; Li, D. Rapid detection of formaldehyde concentration in food on a polydimethylsiloxane (PDMS) microfluidic chip. *Food Chem.* **2009**, *114*, 1079–1082. [CrossRef]
21. Fu, L.; Wang, Y.; Liu, C. An integrated microfluidic chip for formaldehyde analysis in Chinese herbs. *Chem. Eng. J.* **2014**, *244*, 422–428. [CrossRef]
22. Pang, X.; Lewis, A.C. A microfluidic lab-on-chip derivatisation technique for the measurement of gas phase formaldehyde. *Anal. Methods* **2012**, *4*, 2013–2020. [CrossRef]
23. Zhang, X.; Sun, Y.; Fan, Y.; Liu, Z.; Zeng, Z.; Zhao, H.; Wang, X.; Xu, J. Effects of organotin halide perovskite and Pt nanoparticles in SnO_2 - based sensing materials on the detection of formaldehyde. *J. Mater. Sci. Mater. Electron.* **2019**, *30*, 20624–20637. [CrossRef]
24. Baldelli, A.; Jeronimo, M.; Tinney, M.; Bartlett, K. Real - time measurements of formaldehyde emissions in a gross anatomy laboratory. *SN Appl. Sci.* **2020**, *2*, 769. [CrossRef]

25. Ryu, G.; Huang, J.; Hofmann, O.; Walshe, C.A.; Sze, J.Y.Y.; McClean, G.D.; Mosley, A.; Rattle, S.J.; Demello, J.C.; Demello, A.J.; et al. Highly sensitive fluorescence detection system for microfluidic lab-on-a-chip. *Lab Chip* **2011**, *11*, 1664–1670. [CrossRef] [PubMed]
26. Babikian, S.; Li, G.P.; Bachman, M. A Digital Signal Processing-Assisted Microfluidic PCB for On-Chip Fluorescence Detection. *IEEE Trans. Components, Packag. Manuf. Technol.* **2017**, *7*, 846–854. [CrossRef]
27. Christopoulos, T.K.; Diamandis, E. *FLUORESCENCE IMMUNOASSAYS*; ACADEMIC PRESS, INC.: Cambridge, MA, USA, 1996.
28. Wei, L.; Yan, W.; Ho, D. Recent advances in fluorescence lifetime analytical microsystems: Contact optics and CMOS time-resolved electronics. *Sensors (Switzerland)* **2017**, *17*, 2800. [CrossRef]
29. Galievsky, V.A.; Stasheuski, A.S.; Krylov, S.N. Improvement of LOD in Fluorescence Detection with Spectrally Nonuniform Background by Optimization of Emission Filtering. *Anal. Chem.* **2017**, *89*, 11122–11128. [CrossRef]
30. Salthammer, T. Photophysical properties of 3,5-diacetyl-1,4-dihydrolutidine in solution: Application to the analysis of formaldehyde. *J. Photochem. Photobiol. A Chem.* **1993**, *74*, 195–201. [CrossRef]
31. Pokhriyal, A.; Lu, M.; Chaudhery, V.; Huang, C.; Schulz, S.; Cunningham, B.T. Photonic crystal enhanced fluorescence using a quartz substrate to reduce limits of detection. *Opt. Express* **2010**, *18*, 24793–24808. [CrossRef]
32. Weber, M.J. *Handbook of Optical Materials*; CRC Press LLC: Boca Raton, FL, USA, 2003; ISBN 0849335124.
33. Măriuţa, D.; Baldas, L.; Brandner, J.; Le Calvé, S.; Barrot, C.; Magaud, P.; Laurien, N. Design, optimization and manufacturing of a miniaturized fluorescence sensing device. In Proceedings of the 3rd MIGRATE International Workshop, Bastia, France, 24–29 June 2018; p. 210421.
34. Yeh, P.; Yeh, N.; Lee, C.H.; Ding, T.J. Applications of LEDs in optical sensors and chemical sensing device for detection of biochemicals, heavy metals, and environmental nutrients. *Renew. Sustain. Energy Rev.* **2017**, *75*, 461–468. [CrossRef]
35. Tanaka, K.; Choi, Y.J.; Moriwaki, Y.; Hizawa, T.; Iwata, T.; Dasai, F.; Kimura, Y.; Takahashi, K.; Sawada, K. Improvements of low-detection-limit filter-free fluorescence sensor developed by charge accumulation operation. *Jpn. J. Appl. Phys.* **2017**, *56*, 2–7. [CrossRef]
36. Choi, Y.J.; Takahashi, K.; Matsuda, M.; Hizawa, T.; Moriwaki, Y.; Dasai, F.; Kimura, Y.; Akita, I.; Iwata, T.; Ishida, M.; et al. Filter-less fluorescence sensor with high separation ability achieved by the suppression of forward-scattered light in silicon. *Jpn. J. Appl. Phys.* **2016**, *55*. [CrossRef]
37. Nakazawa, H.; Yamasaki, K.; Takahashi, K.; Ishida, M.; Sawada, K. A filter-Less multi-Wavelength fluorescence detector. In Proceedings of the 2011 16th International Conference on Solid State Sensors Actuators and Microsystems TRANSDUCERS'11, Beijing, China, 5–9 June 2011; pp. 100–103.
38. Ding, Z.; Xu, C.; Wang, Y.; Pellegrini, G. Ultra-low-light CMOS biosensor complements microfluidics to achieve portable diagnostics. *Procedia Technol.* **2017**, *27*, 39–41. [CrossRef]
39. Massart, D.L.; Auke Dijkstra, L.K. Chapter 6 Sensitivity and limit of detection. In *Evaluation and Optimization of Laboratory Methods and Analytical Procedures*; Elsevier: Amsterdam, The Netherlands, 1978; pp. 143–156.
40. Guduru, S.S.K.; Scotognella, F.; Chiasera, A.; Sreeramulu, V.; Criante, L.; Vishnubhatla, K.C.; Ferrari, M.; Ramponi, R.; Lanzani, G.; Vázquez, R.M. Highly integrated lab-on-a-chip for fluorescence detection. *Opt. Eng.* **2016**, *55*, 097102. [CrossRef]
41. Sunaga, Y.; Haruta, M.; Takehara, H.; Ohta, Y.; Motoyama, M.; Noda, T.; Sasagawa, K.; Tokuda, T.; Ohta, J. Implantable CMOS imaging device with absorption filters for green fluorescence imaging. In Proceedings of the Optical Techniques in Neurosurgery, Neurophotonics, and Optogenetics, San Francisco, CA, USA, 1–4 February 2014; Volume 8928, pp. 5–10.
42. Torres-Mapa, M.L.; Singh, M.; Simon, O.; Mapa, J.L.; Machida, M.; Günther, A.; Roth, B.; Heinemann, D.; Terakawa, M.; Heisterkamp, A. Fabrication of a Monolithic Lab-on-a-Chip Platform with Integrated Hydrogel Waveguides for Chemical Sensing. *Sensors (Switzerland)* **2019**, *19*, 4333. [CrossRef] [PubMed]

© 2020 by the authors. Licensee MDPI, Basel, Switzerland. This article is an open access article distributed under the terms and conditions of the Creative Commons Attribution (CC BY) license (http://creativecommons.org/licenses/by/4.0/).

MDPI\
St. Alban-Anlage 66\
4052 Basel\
Switzerland\
Tel. +41 61 683 77 34\
Fax +41 61 302 89 18\
www.mdpi.com

Micromachines Editorial Office\
E-mail: micromachines@mdpi.com\
www.mdpi.com/journal/micromachines

www.ingramcontent.com/pod-product-compliance
Lightning Source LLC
LaVergne TN
LVHW070152120526
838202LV00013BA/918